Underwater Domain Awareness

This book presents a comprehensive analysis of the emerging underwater challenges facing India in the Indian Ocean region. With major economic powers like China, the United States, and Russia modernizing their submarine fleets and building advanced unmanned underwater vessels to enhance surveillance capabilities, the competition in the Indo-Pacific underwater domain has intensified.

This book

- Focuses on the issues of detecting, tracking, and classifying submarines/ underwater drones in the Indian Ocean.
- Examines the Indian Navy's present anti-submarine warfare (ASW) capabilities in combating underwater threats and discusses the scope for inter-agency, inter-departmental cooperation framework to monitor the undersea activity in the region.
- Studies the naval composition and strengths of India and other countries in the neighbourhood and reviews maritime domain awareness practices employed by leading navies including NATO for submarine detection.
- Assesses the technology development efforts to deal with these challenges and brings out recommendations.

An expert study of undersea surveillance, this book will be indispensable to students and researchers of military and strategic studies, defence studies, critical security, conflict resolution, intelligence studies, and security studies. It will also be of interest to governments, naval establishments, think tanks, and public policy institutes.

Prakash Pannerselvam is Assistant Professor with the International Strategic and Security Studies Programme, NIAS, India. He holds a PhD degree in Japanese studies from the School of International Studies, Jawaharlal Nehru University, India. He was the first visiting fellow to

Japan Maritime Self-Defence Force (JMSDF) Staff College, Tokyo. He is a recipient of the Japan Foundation Fellow (2011) and Okita Memorial Fellowship (2008). He was also a trained merchant mariner with Poompuhar Shipping Corporation Ltd. Prior to joining NIAS, he was working with Confederation of Indian Industry (CII) in the Defence and Aerospace Sector.

Rajaram Nagappa, a specialist in aerospace propulsion, has worked extensively in the design and development of solid propellant rockets. His interests are in missile technology and space weaponization. He has served at the Vikram Sarabhai Space Centre, ISRO, as its Associate Director, and later was Pandalai Memorial Chair Professor at Anna University, Chennai, India. He has also taught at Technion-Israel Institute of Technology, Israel. He is a recipient of the Astronautical Society of India Award, Distinguished Alumnus Award of the Madras Institute of Technology, DRDO's Agni Award for Excellence in Self Reliance, Certificate of Appreciation of the International Astronautical Federation, and the Honorary Fellowship of the High Energy Materials Society of India. His recent work includes an assessment of the Pakistani cruise missiles and an assessment of the Iranian satellite launch vehicle *Safir*. He has also traced the development of fighter aircraft in China as a part of a study on China's science and technology capability.

R.N. Ganesh (Retd. Vice Admiral) has commanded a diesel submarine, a nuclear submarine, and the aircraft carrier INS *Vikrant*. He has served as the Flag Officer Submarines, Flag Officer Commanding Western Fleet, and Commander Andaman and Nicobar Islands. His other assignments include Director General, Indian Coast Guard, and Commanding-in-Chief, Southern Naval Command. His last appointment was as the Director General of the Indian nuclear submarine programme, which he continued to head after retirement from active service till 2004. He has studied at the Defence Services Staff College, Wellington, the National Defence College, New Delhi, India, and the Adm. Makarov Pacific Fleet Higher Naval Academy, Vladivostok, Russia. His interests include maritime security and strategic affairs and maritime security developments. He is currently collaborating with ISSSP on the Chinese Anti-ship Ballistic Missile capability studies.

Underwater Domain Awareness

Case for India

Prakash Panneerselvam, Rajaram Nagappa, and R.N. Ganesh

LONDON AND NEW YORK

First published 2022
by Routledge
4 Park Square, Milton Park, Abingdon, Oxon OX14 4RN

and by Routledge
605 Third Avenue, New York, NY 10158

Routledge is an imprint of the Taylor & Francis Group, an informa business

British Library Cataloguing-in-Publication Data
A catalogue record for this book is available from the British Library

Library of Congress Cataloging-in-Publication Data
A catalog record for this book has been requested

ISBN: 978-1-032-19268-0 (hbk)
ISBN: 978-1-032-28757-7 (pbk)
ISBN: 978-1-003-29838-0 (ebk)

DOI: 10.4324/9781003298380

Typeset in Times New Roman
by Apex CoVantage, LLC

Contents

List of Tables   vi
Acknowledgements   vii
Abbreviations   viii

1   Introduction: Underwater Challenges in the
    Indian Ocean   1

2   Undersea Threat Environment in the Indian Ocean   10

3   The Evolution of Undersea Surveillance   37

4   Emerging Era in Anti-Submarine Warfare   57

5   Assessing India's ASW Capability in the Indian Ocean   71

6   Conclusion: Unfolding Security Scenario in the
    Indo-Pacific Ocean   91

    Index   100

Tables

1.1 Key Indicators of Shipping Channels in the Indian
Ocean Region   2
2.1 Status of Chinese Submarine Fleet   13
2.2 Noise Level Estimation of Chinese Submarines   14
2.3 PLA Navy's Submarine Launch Anti-Ship Cruise Missile   16
2.4 Chinese Investment in Port Development in the Indian Ocean   18
2.5 PLA Navy's Presence in the Indian Ocean   19
2.6 Pakistan's Present and Future Underwater Capability   23
2.7 Submarines in the Indian Ocean Region   24
3.1 Principle Surface Vessels in ASW Role   49
3.2 Principal ASW Helicopter and Maritime Patrol Aircraft   50
4.1 Full-Spectrum ASW   60
5.1 List of ASW Squadron and Dornier Squadron   73
5.2 Indian Procurement of ASW Aircraft from the United
States and Russia, 2000–2019   73
5.3 Long-Range Maritime Patrol and ASW Aircraft   74
5.4 Indian Navy's Principal Surface and Subsurface Platforms
for ASW Operations   79
5.5 Underwater Surveillance System Developed by DRDO   83
5.6 DRDO's Sonar Systems in Service with Indian Navy   85
6.1 Indian Navy's Key Objectives, Strategy, and Role of UDA   95
6.2 Strategic Partners: Monitoring Chokepoints in
the Indian Ocean   99

Acknowledgements

Working on this monograph has been an interesting and enriching experience. Thanks to a visit arranged by Vice Admiral RN Ganesh (Retired) to the Southern Naval Command, the authors could garner some practical aspects of the intricacies of the underwater domain. Our thanks to the Officers of the Anti-Submarine Warfare School of the Indian Navy for sparing their time and sharing their knowledge. Our thanks to the Director and Scientists of the Naval Physical and Oceanographic Laboratory, Kochi, for sharing information on the status of sonar sensor development in India. The visit to the National Centre for Ocean Information Services (INCOIS), Hyderabad, provided useful information on tsunami sensors and the tsunami warning network operated by India. Our thanks to the Director and Scientists of INCOIS for sparing their time and sharing useful information.

We are grateful to Prof. Srikumar Pullat, Head of International Strategic and Security Studies Programme (ISSSP), for his constant support and encouragement, which has made this publication possible. We would like to extend our thanks to Prof. Suba Chandran, Dean of School of Conflict and Security Studies at NIAS for his support. Head Administration and his colleagues have provided timely administrative support under the trying situation of the pandemic and our sincere thanks go out to them.

We would like to express our gratitude to the Dr Shailesh Nayak, Director, National Institute of Advanced Studies (NIAS) for his support and encouragement to the work. Maritime security and ecology is a subject close to his heart and the authors have benefitted from discussion with him on these topics including aspects of Blue Economy.

Our thanks to the anonymous reviewers of the monograph manuscript. Their comments and suggestions have helped in adding content and rigour to our effort. Our thanks to Routledge – especially to Ms Shoma Choudhury – for all the effort in making this publication possible.

Abbreviations

AI	Artificial Intelligence
A2AD	Anti-Access/Anti-Denial
ACTAS	Active Towed Array Sonar
ACTUV	ASW Continuous Trail Unmanned Vessel
AIP	Air-Independent Propulsion
ALTAS	Advanced Light Towed Array Sonar
AMTI	Asia Maritime Transparency Initiative
APSOH	Advanced Panoramic Sonar Hull Mounted
ASCMs	Anti-Ship Cruise Missiles
ASS	Autonomous Seabed Station
ASW	Anti-Submarine Warfare
ATDS	Advanced Torpedo Defence System
AUKUS	Australia, United Kingdom, The United States
BEL	Bharat Electronics Limited
BRI	Belt and Road Initiative
CCS	Cabinet Committee on Security
CDAA	Circularly Disposed Antenna Array
CETC	China Electronics Technology Group Corporation
CMRE	Centre for Maritime Research and Experimentation
CMSI	China Maritime Studies Institute
CN3	Communications/Navigation Network Node
COMINT	Communication Intelligence
COMRA	China Ocean Mineral Resources R & D Association
COP	Common Operational Picture
CSI	Container Security Initiative
CSIC	China Shipbuilding Industry Corporation
CSTC	China Shipbuilding Trading Company
DAC	Defence Acquisition Committee
DARPA	Defence Advanced Research Projects Agency
DASH	Distributed Agile Submarine Hunting

DNS	Distributed Netted Systems
DPRK	Democratic People's Republic of Korea
DRDO	Defence Research & Development Organisation
DSRV	Deep Submergence Rescue Vehicles
EEZ	Exclusive Economic Zone
ELINT	Electronic intelligence
ESM	Electromagnetic Support Measure
GIUK Gap	Greenland, Iceland, England Gap
GUGI	*GlavnoyeUpravlenieGlubokovodskIssledovanii*
HA/DR	Humanitarian Assistance/Disaster Relief
HAL	Hindustan Aeronautical Limited
HF/DF	High-frequency direction finding
HUMSA	Hull-Mounted Sonar Advanced
ICBM	Intercontinental Ballistic Missile
IIFP	Floating Integrated Information Platform
INDUS-NET	Indian Distributed Underwater Surveillance Network
IOR	Indian Ocean Region
IRBIS	Island Reef-Based Integrated Information System
ISPR	Inter-Services Publication Relations
ISR	Intelligence, Surveillance, and Reconnaissance
ISRO	Indian Space Research Organisation
IUSS	Integrated Undersea Surveillance System
IUU	Illegal, Unregulated and Unreported Fishing
JMSDF	Japan Maritime Self-Defence Force
JNB	Jinnah Naval Base
JOED	JMSDF Ocean Surveillance Information System
LEMOA	Logistic Exchange of Memorandum of Agreement
LFA	Low-Frequency Active Sonar
LFDS	Low-Frequency Dunking Sonar
LFVDS	Low-Frequency Variable-Depth Sonar
LOFAR	Low-Frequency Analysis and Raging
LTTE	The Liberation Tiger of Tamil Eelam
MAD	Magnetic Anomaly Detection
MAITRI	Marine & Allied Interdisciplinary Training and Research Initiative
MAWS	Missile Approach Warning Systems
MCM	Mine Countermeasures
MDA	Maritime Domain Awareness
MIT	Massachusetts Institute of Technology
MOSFET	Metal-Oxide-Semiconductor Field-Effect Transistor
MPA	Maritime Patrol Aircraft
MSR	Maritime Silk Road

NATO	North Atlantic Treaty Organisation
NPOL	Naval Physical and Oceanographic Laboratory
NS&M	Naval Systems and Materials
NSB	North Strategic Bastion
ODIS	Ocean Data and Information System
ORBAT	Overall Order of Battle
PAFC	Phosphoric Acid Fuel Cell
PLA Navy	People Liberation Army Navy
PMS	Presence and Surveillance Missions
PTAS	Passive Towed Array Sonar
QUAD	Quadrilateral Security Dialogue
RDSS	Rapidly Deployable Surveillance System
ROK	Republic of Korea
SACLANT	Supreme Allied Commander Atlantic
SAGAR	Safety And Growth of All in the Region
SHARK	Submarine Hold at RisK
SIGINT	Signal Intelligence
SL	Source Level
SLBM	Submarine-Launched Ballistic Missile
SLCM	Submarine-Launched Cruise Missile
SLOCs	Sea Lines of Communication
SOSUS	Sound Surveillance System
SPAWAR	Space and Naval Warfare
SSBN	Ballistic Missile Submarine
SSK	Diesel-Electric Submarine
SSN	Nuclear-Powered Submarine
SST	Sea Surface Temperature
SURTASS	Surveillance Towed Array Sensor Systems
TASS	Towed Array Sonar System
TCS	Time Critical Strike
TNI-AL	*Tentar Nasional Indonesia-Angkatan Laut*
ToT	Transfer of Technology
TRAPS	The Transformational Reliable Acoustic Path System
UDA	Underwater Domain Awareness
UNCLOS	United Nation Convention on Law of the Sea
USNI	United States Naval Institute
USSR	Union of Soviet Socialist Republics
USV	Unmanned Surface Vessels
UUSs	Unmanned underwater systems
UUV	Unmanned Underwater Vessel
UWACS TRITON	Underwater Wireless Acoustic Communication System
VDS	Variable-Depth Sonars
VLF	Very-Low Frequency

1 Introduction

Underwater Challenges in the Indian Ocean

The Indian Ocean occupies a geostrategically important place in global politics. The geographical location and combination of various factors like the presence of major economies, nuclear power countries, and strategic sea lines of communication (SLOCs) connecting energy and trade routes across the Indian Ocean make this region the fulcrum of global politics and security. Major geopolitical challenges to the Indian Ocean emerge from two considerations. Firstly, the maritime nations depend on the Indian Ocean for trade, energy, and commerce prospects. Secondly, the regional and extra-regional powers view the Indian Ocean as a 'strategic space' coming under the maxim, *whoever controls the space, controls the region*. As a maritime power, India shoulders a huge responsibility in safeguarding the crucial sea lanes of the Indian Ocean. Strategic experts claim that India's national security can be best guaranteed only by expanding India's security perimeter and achieving a position of influence in the Indian Ocean. India's strategic space extends from the western shores of Australia to the eastern coast of Africa, which includes the area between the Strait of Malacca in the east to the Strait of Hormuz in the west. To safeguard and control this strategic space demands an effective situational awareness in place. Technology plays an important role in the selection and deployment of sensors for exercising situational awareness on the sea surface and underwater domains. While one can deploy sensors on the land, sea, air, and space for monitoring surface traffic, one needs a different strategy for monitoring the movement of underwater platforms. India has adequate capability and capacity for the surveillance of surface movements, and it needs to prioritize the development of a comprehensive underwater domain awareness (UDA) strategy (Gokhale, 2020), which is the need of the hour.

Geopolitical Features of the Indian Ocean

The Indian Ocean is the third largest ocean in the world. It covers an area of about 78 million square kilometres, comprising about 14 percent of the

DOI: 10.4324/9781003298380-1

earth's land surface and 20 percent of the earth's water surface (Central Intelligence Agency, n.d.). It includes, amongst others, the Red Sea, the Persian Gulf, the Arabian Sea, the Bay of Bengal, and the Andaman Sea. The Indian Ocean is surrounded by land on three sides – on the west by the African continent, the north by Southwest Asia, and the east by Western Australia. As a major sea lane, it connects Asia and Africa with Europe and America. Plush with living and non-living resources, from marine biodiversity to oil and natural gas, the Indian Ocean is economically crucial to Africa, Asia, and Australia/Oceania – the three continents bordering it, and the world at large. The Indian Ocean is a rich energy source with the oil fields of the Gulf, Indonesia, and Brunei having significant share of the world's energy basket.

The Indian Ocean is a critical waterway for global trade and commerce. It is also one of the busiest maritime traffic routes (Davies, 2019). About half of the world's containerized cargo, one-third of its bulk cargo, and two-thirds of its oil shipment pass through the Indian Ocean every year (Bhattacharjee, 2020). The preeminent status of the Indian Ocean as an international SLOC remains well established. Therefore, the security of chokepoint is recognized more than ever. With the globalization of the world economy and the corresponding dependence of a greater number of nations on foreign trade, the Indian Ocean has assumed even greater importance, serving over one-third of the world's seaborne trade. Table 1.1 indicates the major

Table 1.1 Key Indicators of Shipping Channels in the Indian Ocean Region

Strait	Water depth (m)	Width (Km)	Vessel traffic (thousands of ships/year)	Oil throughput (10,000 barrels/day)	Connected sea routes
Malacca Strait	25	37	8	1,520	Arabian Sea, South China Sea
Sunda Strait	20	24	3	-	Java Sea, Indian Ocean
Lombok Strait	150	11.5	Less than thousand ships	-	Java Sea, Indian Ocean
Strait of Hormuz	10.5	56	4	1,700	Persian Gulf, Arabian Sea
Bab al-Mandab Strait	150	26	2	380	Red Sea, Mediterranean Sea, and Arabian Peninsula
Suez Canal	22.5	34	2	320	The Mediterranean Sea, Arabian Peninsula

Source: Compiled by author from (Numbers of Ships Reporting Under STRAITREP, 2021), (ERIA, 2016).

chokepoints and maritime traffic around these straits connecting the Indian Ocean with the rest of the world.

Since a large volume of international maritime trade and energy transits through the crucial sea lanes of the Indian Ocean, this route is a lifeline for global economies, and disruption to these sea lanes will practically affect the growth and economy of many nations. Similarly, the chokepoints in Southeast Asia – mainly the *Malacca*, *Sunda* and *Lombok* Straits – practically affect the economic vitality of the Indo-Pacific region (see Table 1.1). The sheer volume of products and energy goods transiting these straits is almost half of the world's merchant fleet capacity and one-third of the world's ships (US Energy Information Administration, 2017). The sea lanes of the Indian Ocean and the chokepoints in Southeast Asia have great geostrategic importance for India, the United States, Europe, China, Japan, South Korea, Taiwan, and other Southeast Asian countries. Given the fact that the northern Indian Ocean is a hub of maritime activity, it has significant influence on the economic and security interests of the countries in this region.

The situation is volatile in some regions bordering the Indian Ocean. Internal conflicts, terrorism, and state of lawlessness in some of the countries, especially in Somalia, Yemen, and Syria, have turned the region into a hotbed of piracy, drug smuggling, gun-running, and human trafficking. Issues like Illegal, Unregulated and Unreported (IUU) Fishing have aggravated the security scenario. For example, the narcotics trafficking from Pakistan's coast to other parts of the Indian Ocean is weakening state functionary and leading to major law and order situations in the littorals (Panneerselvam, 2021). These factors not only add instability to the region but also pose a serious security risk to merchant vessels passing through the region. The hijacking ships by Somali pirates for ransom between 2008 and 2014 is a good example to show how illegal activity can disrupt global energy and trade.

In response to maritime piracy in the Indian Ocean, navies from the United States, the UK, the European Union, China, and Japan have deployed their naval ships for counter-piracy activity in the western Indian Ocean. China at times has deployed two warships in the region to protect the SLOCs; later, it developed a naval base in Djibouti to sustain its maritime operations in the region. The full-fledged Chinese naval base is equipped to support the PLA Navy's activity in the Indian Ocean and Africa. Japan too has established a naval base in Djibouti to expand Japan Maritime Self-Defence Force (JMSDF) activities in the region. Not to mention, the US Navy, which has an important stake in the region and has a naval base in Diego Garcia, a British Indian Ocean territory, to actively monitor and support the US forces in the region. Through the Belt and Road Initiative (BRI) and Maritime Silk Road (MSR), China is reinforcing its presence in the region and views the Indian Ocean and the Pacific Ocean as critical pathways to achieve energy and trade security. Therefore, maritime domain awareness (MDA) has become an essential requirement for the People Liberation Army (PLA) navy from a security perspective. India is wary of Chinese intentions in the region and is keeping up with its modernization programme by building surface and subsurface naval capabilities to deter Chinese provocation and to improve underwater awareness capability in the region. Issues in the Indian Ocean pertaining to tropical weather conditions, seawater salinity, uncertainties due to biological noise, and underwater radiant noise emitted by marine vessels further complicate the underwater scenario in this region. To assess the subsurface security threat in the Indian Ocean, an understanding of the hydrographic and bathymetric condition of the Indian Ocean would be useful to navigate through the impending challenges faced by naval planners.

Hydrographic and Bathymetric Condition of the Indian Ocean: An Overview

For UDA, the principal sensing mode is acoustic, and the speed of sound in the water medium is affected by the chemical and physical properties of the sea water. The waters of the oceans differ in these properties. Indian Ocean hydrology is distinctly different from that of other oceans. Unlike the Pacific and Atlantic Oceans, the Indian Ocean is closed in the northern subtropics, blocking the effect of the equatorial current system for the spread of water masses in the thermocline (Tomczak & Godfrey, 1994). The flow of freshwater from Asian rivers during the monsoon season into the Indian Ocean, the summer flood water from Ganges and Brahmaputra into Bay of Bengal in the east, the Indus water into Arabian Sea, and the Irrawady and Salweeen Rivers into Andaman Sea strongly influences Indian Ocean's water properties (Tomczak & Godfrey, 1994). The tropical monsoon condition of the region influences the regional oceanographic circulation. The Bay of Bengal receives a large flow of fresh water from the Ganges River and precipitates; the Arabian Sea in contrast loses freshwater to the atmosphere (Gordon, n.d.). Arnold L. Gordon, from the Lamont-Doherty Earth Observatory, found in his research that the Bay of Bengal's complex and energetic mesoscale plays a central role in 'closing the regional scale heat and freshwater budgets both laterally and vertically, affecting the sea surface temperature and air-sea interaction and pycnocline structure' (Gordon, n.d.). According to oceanographers, the sea surface temperature (SST) of the entire northern Indian Ocean appears to be a continuation of the western Pacific, which is generally regarded as the warmest region of the open ocean (Tomczak & Godfrey, 1994). The temperature and salinity of the water are greatly influenced by wind, water current, monsoon seasons, etc. These physical and chemical properties of the Indian Ocean define its basic characteristics and influence its acoustic response. UDA, therefore, makes it imperative to understand the peculiarities of acoustic propagation in the Indian Ocean waters and determine its characteristics.

Propagation of Sound in the Indian Ocean

The study of acoustical oceanography describes the ocean as an acoustic medium. It primarily observes the oceanic property and acoustic behaviour of underwater noise, vibration, etc. According to Kuperman, sound propagation in the ocean is 'governed by the spatial structure of the sound speed and sound speed in the ocean is a function of temperature, salinity, and ambient pressure' (Kuperman, 2001). The sound speed profiles in deep water show the greatest variability near the surface, says Kuperman. During hot days, the temperature increases near the surface and hence the sound speed increases towards the sea surface as well. An important aspect of sound propagation in the ocean medium due to its layered structure is refraction, since according to Snell's law, the angle of refraction is dependent on the instantaneous speed of sound in the layer (Burdic, 1984). Other factors include temperature, depth, and other oceanic properties. When the depth of the ocean is not sufficient, the sound signal interaction with the bottom will result in high attenuations, thereby reducing the sonar range. There are multiple factors that affect sound propagation in oceans; therefore, one can conclude by saying that deep water is conducive for long-range propagation with minimal distortion. In coastal waters or shallow waters, sound propagation is greatly affected by the sea bottom. From the earlier discussions, it is very clear that the propagation of sound in deep waters is very conducive to long-range propagation with minimal distortions and attenuations. However, in littoral waters, the propagation is highly restricted, due to multiple bounces from the sea surface and the sea bottom.

Many oceanographic and naval experts have opined that oceanic property in the Indian Ocean presents difficult sonar conditions. In the Indian Ocean, natural sources of sound include geophysical activities like surface wind, underwater volcanoes, and earthquakes; and biological sources include whale songs, dolphin clicks, and fish vocalizations. Anthropogenic noise originates from human-driven activities like shipping, geophysical surveys, oil and gas exploration, dredging, and sonar transmissions. The classical ambient noise plot by Wenz GM during World War II gives a clear segregation of the sources based on their frequency bands (Wenz, 1962). Extremely low frequencies (<1 KHz) are dominated by distant shipping followed by wind-generated noise up to 15 kHz, yielding a Wenz curve with clear domination of the shipping noise in the low-frequency region (Wenz, 1962).

According to research carried out by the Pennsylvania State University, 'the Indian Ocean sound floor has increased over the past decade, and the greatest increase occurred in recordings from the south side of Diego Garcia in the 85–105Hz range and shipping noise is a large contributor to the increase' (Miksis-Olds, Bradley, & Niu, 2013). Increase in shipping over the last decade has greatly contributed to ambient noise in the Indian Ocean. At the same time, regular factors like wind speed, seismic activity, and the population of blue whales also add to the ambient noise in the Indian Ocean. Therefore, multiple measurement parameters are needed to fully understand the changing acoustic conditions in the Indian Ocean. Greater investment in India's naval hydrography and other associated disciplines will be necessary for India to study the Indian Ocean's acoustic noise properties.

Focus on Underwater Domain Awareness: What Are the Challenges?

Intelligence, Surveillance, and Reconnaissance (ISR) is a necessary 24×7 function to safeguard India's interests in territorial waters, Exclusive Economic Zone (EEZ), and the Indian Ocean. The Indian Navy and the Indian Coast Guard are tasked with the functions of maintaining the security of waters in general circumstances as well as of planning and executing necessary strategies and operations in hostile or war-like conditions. MDA is a key perquisite. As indicated earlier, India's MDA capability relating to surface vessels is at an acceptable level. The underwater domain awareness needs attention and needs to be strengthened – both in capability and in capacity.

India's primary maritime security is built around its coastal and littoral waters. The definition of littoral waters according to the US Naval Doctrine Publication 1 (Navy, March 2010) consists of two segments:

1. Seaward: the area from the open ocean to the shore, which must be controlled to support operations ashore.
2. Landward: the area inland from the shore that can be supported and defended directly from the sea.

In naval operations, the land mass adjacent to the oceans within direct control of and vulnerable to the striking power of sea-based forces is identified. Generally, the areas referred earlier fall in the coastal regions, wherein the water depths are less than 200 m or 50 nautical miles away from the coast.

Shallow waters pose challenges in detecting, identifying, and engaging the adversary's naval assets. In littoral waters, with boundaries framed by the surface and bottom, the typical depth-to-wavelength ratio is about

10–100. That ratio makes the propagation of acoustic waves rather complicated. As a result of this, the environmental and sound characteristics of the littoral region favour a defensive posture.

Given the present security scenario, it becomes necessary for India to build its detection capability in both coastal waters and the high seas to detect threats. In this domain, India's present ASW capabilities require enhancement. Presently, the Indian Navy operates a fleet of Boeing Poseidon-8I aircraft for ASW purposes and will be augmenting the fleet by the addition of P-8I long-range maritime patrol aircraft. In addition, the navy also operates helicopters like Westland Sea king, Sikorsky Sea King, Kamov 28, and Kamov 31 in ASW roles. Indian Navy surface vessels are equipped with indigenously built Hull-Mounted Sonar Advanced (HUMSA) and Passive Towed Array Sonar (PTAS) sourced from Thales. The Indian Navy surface vessels are equipped with advanced systems for anti-submarine operations in its water.

The ASW sonar technology has grown exponentially in some of the advanced countries. The Active Towed Array Sonar (ACTAS) operates at low frequencies and is deployable at varied depths to achieve long-range submarine detection. The first six units of the ACTAS system imported from Germany were fitted in Talwar-class frigates and Delhi-class destroyers. Additional requirements will be met by the Bharat Electronics Limited (BEL) manufacturing under Transfer of Technology (ToT). The Indian Navy is also planning to equip Kamorta-class corvettes, Shivalik-class frigates, and Kolkata-class destroyers with the ACTAS system. These systems, while adding to domain awareness, are not adequate to meet the growing underwater challenges in the region.

Moreover, undersea warfare can be expected to undergo significant changes with the incorporation of new technologies and deployment of unmanned underwater vessels (UUVs). The advances in battery and fuel cell technology are expected to enable the endurance of UUVs and can operate deep in the ocean, thereby changing the nature of Anti-Submarine Warfare (ASW). The United States, Russia, NATO countries, and China are in the process of developing the next level of autonomous underwater drone with artificial intelligence (AI) features to detect, track, and identify the quietest enemy submarine, even in harsh sea conditions. The emerging ASW technologies appropriate to the threat scenario need to be identified and taken up for development, procurement, or co-development.

Non-acoustic detection techniques like Magnetic Anomaly Detection (MAD), hydrodynamics, laser detection, and Electromagnetic Support Measure (ESM) are showing progress in submarine detection. The MAD systems are currently deployed in ASW aircraft and helicopters. Technology breakthroughs with laser and LED systems hold promise and are being pursued vigorously. With the latest advancements in lasers and LEDs and advancements in the field of big data computing and analyses, even faint signals emanating from submarines can be detected. This calls for a focused research on the use of big data in the field of submarine detection techniques. This technique aids submarine detection by comparing expected ambient noise from marine life, waves, and seismic events from the measured noise fields, to possibly identify noise signals getting reflected off a submarine or hull (Frey, Gagnon, & Tart, 1996). As the computing power is rapidly increasing and becoming more portable, it can add advantage to both acoustic and non-acoustic modes of detection.

From India's security considerations, it is imperative to possess MDA with respect to adversary surface and subsurface assets at any given time. This calls for the development/procurement and deployment of appropriate sensors and platforms to monitor adversary sea assets from air, surface, subsurface/ocean bed, and space on a continuous basis.

Objective of the Study

The detailed introduction in the earlier sections has brought out the major challenges for India, to detect, track, and identify submarines in the Indian Ocean. The new class of submarines fielded and planned to be fielded by the PLA Navy are faster, stealthier, and quieter, as well as capable of operating in shallow waters. Complicating the issue further are the growing presence of diesel submarines by littorals and the introduction of UUV platforms. India needs to develop a long-term strategy for wide-area surface and underwater surveillance, keeping the peculiarities of the salinity of the waters in mind. This study plans to examine and explore the naval forces operating in the Indian Ocean Region (IOR), their force structure, international MDA/UDA practices, threat scenarios, and the possible MDA/UDA approach India should adopt.

Bibliography

Bhattacharjee, S. (2020, June 24). Beyond Galwan Fastness: Energy Security in Indian Ocean May Be a Challenge. *Business Standard*. Retrieved from: www.business-standard.com/article/economy-policy/beyond-galwan-fastness-energy-security-in-indian-ocean-may-be-a-challenge-120062401961_1.html (Accessed on September 12, 2021).

Burdic, W. S. (1984). *Underwater Acoustic System Analysis*. Englewood Cliffs: Prentice Hall.

Central Intelligence Agency. (n.d.). *The World Factbook*. Retrieved from: www.cia.gov/the-world-factbook/oceans/indian-ocean/#environment (Accessed on April 12, 2021).

Davies, R. (2019, September 5). A Game of Risk: The Indian Ocean's Most Strategically Important Ports. *Ship Technology*. Retrieved from: www.ship-technology.com/features/a-game-of-risk-the-indian-oceans-most-strategically-important-ports/ (Accessed on September 11, 2021).

ERIA. (2016). *Sea Lane Security of Oil and Liquefied Natural Gas in the EAS Region*. Jakarta: ERIA.

Frey, A. R., Gagnon, J. R., & Tart, J. (1996). Detection of a Silent Submarine From Ambient Noise Field Fluctuations. *UMAP Journal*, 17(3).

Gokhale, V. (2020, June 23). There Is a Pressing Need for India to Develop a Comprehensive Underwater Domain Awareness Strategy. *The Indian Express*. Retrieved from: https://indianexpress.com/article/opinion/columns/india-china-border-dispute-galwan-sea-route-6471403/ (Accessed on October 23, 2021).

Gordon, A. L. (n.d.). *Bay of Bengal Surface and Thermocline and the Arabian Sea*. Office of Naval Research. Retrieved from: www.onr.navy.mil/reports/FY14/pogordo2.pdf (Accessed on October 7, 2021).

Kuperman, W. (2001). Acoustics, Deep Ocean. In J. H. Steele (ed.), *Encyclopedia of Ocean Sciences* (pp. 61–72). New York: Elsevier.

Miksis-Olds, J. L., Bradley, D., & Niu, X. M. (2013). Decadal Trends in Indian Ocean Ambient Sound. *The Journal of the Acoustical Society of America*, 3464–3475.

Navy, U. (March 2010). Naval Doctrine Publication 1. United States Government.

Numbers of Ships Reporting Under STRAITREP. (2021, October). *Marine Department Malaysia*. Retrieved from: www.marine.gov.my/jlm/admin/assets/uploads/images/contents/20211108103049-1aa77-bilangan-kapal-melapor-di-bawah-straitrep-sehingga-oktober2021.pdf (Accessed on November 12, 2021).

Panneerselvam, P. (2021). Maritime Narcotics Trafficking in the Western Indian Ocean: Threat to Regional Maritime Security. *Maritime Affairs: Journal of the National Maritime Foundation of India*, 110–115.

Tomczak, M., & Godfrey, J. (1994). *Regional Oceanography: An Introduction*. Pergamon: Pergamon.

US Energy Information Administration. (2017, August 11). Retrieved from: www.eia.gov/todayinenergy/detail.php?id=32452#:~:text=Nearly%20one%2Dthird%20of%20the,after%20the%20Strait%20of%20Hormuz(Accessed on October 11, 2020).

Wenz, G. M. (1962). Acoustic Ambient Noise in the Ocean: Spectra and Sources. *The Journal of the Acoustical Society of America*, 1936–1956.

2 Undersea Threat Environment in the Indian Ocean

A principal threat to India emerges from hostile submarines operating close to the Indian coast. An unlocated submarine in the Indian Ocean is a security threat that can create major military, diplomatic, and economic consequences for India. Unlike other surface vessels, submarines have a very limited peacetime role, and they will almost always be perceived as a threat by the opponent. Indian naval analyst Abhijit Singh claims that 'Submarines are quintessential war platforms – vastly potent, but also highly provocative, and incapable of subtle signaling' (Singh, 2021). This statement is true because even midget-class submarines – which are limited in range and capabilities – can trigger extreme reactions from the opponents. The western navies have realized the need for better situation awareness during peacetime, particularly to monitor and protect Exclusive Economic Zones (EEZs) and territorial waters from enemy activities. The navy's traditional role of gathering surveillance, reconnaissance, and intelligence on the enemy's activity is also a fundamental aspect driving the underwater domain awareness concepts. Moreover, the maritime domain has become increasingly more complex as the underwater environment is cluttered and chaotic, and defeating stealthy enemies is becoming a major challenge.

Further adding to the complex underwater maritime environment are the latest developments in weapon technologies. The major naval forces are actively pursuing 'sea-denial' to keep other navies at bay. The proliferation of such technologies to the other countries in the Indo-Pacific region has significantly increased in the recent past, thus adding complexity to the underwater security environment in the Indian Ocean. For example, the submarine-launched anti-ship cruise missiles/land-attack cruise missiles, diesel submarines with air-independent propulsion (AIP) systems, midget-class submarines and unmanned underwater vessels (UUVs)/gliders for maritime intelligence, surveillance, and reconnaissance is adding insecurity to the underwater maritime domain in this region. In addition to

DOI: 10.4324/9781003298380-2

that, the presence of active non-state groups with maritime capabilities in Pakistan may leverage their skills to attack surface and underwater targets in the region. The non-state actors or organized criminals may also use the semi-submersibles or underwater vehicles for smuggling narcotics and gunrunning in the future as technology becomes more easily available in the open market.

For India, the major threat in the past emerged from the Pakistan Navy's submarines. India and Pakistan fought four major wars, of which the 1965 and 1971 wars involved an active naval engagement between the two nations. The PNS *Ghazi*, Pakistan navy's flagship submarine, was actively involved in the 1965 war against India in the Arabian Sea. The Vice Admiral of Pakistan Navy (retired) Taj M Khattak said that 'PNS Ghazi played a pivotal role in blockading the Indian Navy fleet in Mumbai Harbour' (Khattak, 2018). On the eve of ceasefire, on September 23, 1965, Pakistan Radio made a false claim that PNS *Ghazi* sank INS *Brahmaputra*. The Indian Navy refuted such claims, as INS *Brahmaputra* was not attacked or sunk by the PNS *Ghazi* (Gokhale, 2015). In fact, PNS *Ghazi* was kept under pressure by Indian Navy surface ships and aircraft during the course of the war. During the 1971 war, PNS *Ghazi* posed a major challenge to India's sole aircraft carrier INS *Vikrant,* which was enforcing the naval blockade in East Pakistan. The PNS *Ghazi* deployed in a mission to locate and attack the INS *Vikrant* mysteriously sank off the Coast of Visakhapatnam on December 3, 1971. The reason behind the sinking of PNS *Ghazi* still remains unknown. However, the incident has raised serious concern over India's preparedness in the underwater domain to combat such threats in the future. India's threat perception is also changing with the rising security threats to India in this region. India embarked on the armed force modernization plan to strengthen its military capability including a plan to lease a Charlie-1 class nuclear submarine from Soviet Union for training and familiarization purposes. After India's nuclear test in 1998, its nuclear doctrine to build a nuclear triad placed a strong emphasis on the nuclear submarine programme. In 1999, the Cabinet Committee on Security (CCS) approved the Indian Navy's 30-year submarine construction plan to build 24 diesel-electric submarines (SSKs) and 6 Ballistic Missile Submarines (SSBNs). Particularly, after the Mumbai 26/11 terror attacks, India's focus on the maritime domain has substantially increased to watch out against surface and subsurface threats in these waters. Indian Navy Chief Admiral Karambir Singh reported that Pakistan-based Jaish-e-Mohammad is training mujahideen in underwater attacks to target Indian ships and energy assets in Indian waters or abroad (India Today, 2019). In addition, India is facing a new set of challenges in the underwater domain, which requires a detailed assessment of emerging challenges in the Indian Ocean:

1. Chinese submarines in the Indian Ocean
2. Sea-based nuclear capability of Pakistan
3. Proliferation of conventional submarines
4. Emerging unmanned underwater vehicles (UUVs)
5. Threats of submarine incidents or accidents in the Indian Ocean.

The geopolitical environment in the Indian Ocean is fast changing and brings new challenges to India's maritime security. This chapter will largely focus on how the Chinese and Pakistani submarine programmes, as well as the proliferation of conventional submarines, unmanned underwater systems (UUSs), and the threat of submarine accidents – which pose a major subsurface challenge to India.

1. Chinese Submarine Capability and Deployment in the Indian Ocean

Chinese SSBNs and nuclear-powered submarine (SSNs) are important to maintain strategic stability in the region. The Chinese SSBNs/SSN operational strategy is based on a two-pronged approach: The first is to have submarine conduct patrols in the open ocean (whose objective is to primarily rely on their own stealth capabilities when they transit to patrol area and stay undetected during the patrol period). The second approach is to deploy SSBNs to designated areas (termed as bastion) in coastal waters to protect the submarine from the enemy's advanced Anti-Submarine Warfare (ASW) capability (Washington, 2017). However, the two-pronged strategy has its own limitations: Chinese naval capabilities, particularly in building nuclear class submarines and submarine-launched ballistic missiles (SLBMs), required some advancement to fully acquire second-strike deterrence capability against the United States. The 2019 Defence White Paper of China says that the PLA Navy is transiting from 'near-seas protection' role to an 'open seas protection' role, which aligns with the PLA Navy's aim of open-ocean patrol to maintain credible deterrence against the United States. Based on this approach, Chinese sea-based deterrence capability can be understood from two distinct developments:

1. First, China's first nuclear submarine Type-092 class was completed in 1981 and entered service in 1983. Due to major problems, it never reached the level of a successful deterrent. The Type-091 has no combat value and was used only in training future submariners.

2. Second, based on the Type-092 experience, China's naval modernization embarked on the development of newer-class nuclear submarines in the 1990s. The Type-094 Jin-class ballistic missile submarine and the Type-093 Shang-class nuclear attack submarines are the two new variants of the 1990s technology that are currently in operation (see Table 2.1). China is also advancing its research and development in submarine technology and is planning to build newer-class SSBNs/SSNs with indigenous technology.

At present, the *Jin-c*lass SSBNs equipped with JL-2, a submarine launch version of the DF 31, are an advanced intercontinental ballistic missile (ICBM) with an operation range of 7,000 km; it is the main weapon of China's sea-based deterrence. Strategic experts have raised questions about the credibility of this weapon system because it poses a credible threat to the US mainland, China has to deploy its Type-04 SSBN in the Pacific Ocean, to launch the JL-2 missiles. The limited experience of Chinese Type-04 SSBNs and the operational ceiling of JL-2 SLBMs limit the value of this system as a strategic weapon. China is working on a more advanced version of JL-3 SLBMs, which can deliver multiple warheads within the range of 10,000–12,000 km. According to South China Morning Post (Chan, 2021), China inducted a new Type of 094A Jin-class submarine with improved 'hydrokinetic and turbulent systems' making it more suitable to carry the

Table 2.1 Status of Chinese Submarine Fleet

SSBN	Numbers
Jin (Type-094)	6
SSN	
Shang I (Type-093)	2
Shang II (Type-93A)	4
Han (Type-091) (In reserve)	3
SSK	
Kilo (Project 877)	2
Improvised *Kilo* (Project 636)	2
Improvised *Kilo* (Project 636M)	8
Ming (Type-035B)	8
Song (Type-039 G)	12
Yuan (Type-039A)	4
Yuan II (Type-039B)	14
Ming (Type-035 G) (In Reserve)	8
SSB	
Qing (Type-032)	1

Source: Table is based on IISS Military Balance 2021 (Asia, 2021).

more powerful JL-3 SLBMs (Chan, 2021). The future version of Type-05 and Type-06 SSBNs with JL-3 SLBMs would pose credible nuclear deterrence against the United States (Chan, 2021). The Type-093 (Shang class) has significant commonality with Type-094 Jin class. According to Janes Defence, the Type-093 carries submarine lunch version of YJ-18 and CJ-10 anti-ship cruise missiles. Capt. (retd.) Jerry Hendrix, director of the Defense Strategies and Assessments Program at the Center for a New American Security, says that Type-093B is a transition platform between the Type-093 and the forthcoming Type-095 (Majumdar, 2016). The future Chinese submarines may be able to match with the US Navy's Los Angeles-class vessels.

The fundamental challenge for China is building quieter submarines. Based on the US office of Naval Intelligence 2009 report, the Chinese Type-094 class submarine is noisier than the Russian Delta III SSBN (Office of Naval Intelligence, U.N., August 2009). According to Eugene Miasnikov, the Russian Delta III SSBN's noise level was 125–130 db, in natural condition as the Delta III has a 15 percent chance of getting detected at a distance of 30 km by the US Los Angeles-class SSN (Miasnikov, 1995). Wu Riqiang, Chinese scholar and associate professor, School of International Studies, Renmin University of China, has also raised similar concerns about the Chinese submarines being noisier and questioned the survivability of Chinese SSBNs. Prof. Wu also provides the following estimates of noise level in the Chinese submarines, which seems not adequate for China to deploy Type-094 in deterrent patrol in the open oceans.

Chinese military establishments have also realized that the Chinese Type-094 class submarine is noisier and detectable by enemy ASW, making the survivability of SSBNs questionable. According to Chinese scholar Zhao, some of the basic features of the Type-094 submarine is limiting its potential to become a quiet submarine (Bell, 2009)[1]. Despite such limitations and given the ambient noise in the South China Sea (70–96 db), the detection range of Type-094 would be at 50km if its source level (SL) is 140 db, says

Table 2.2 Noise Level Estimation of Chinese Submarines

Submarine	Type	Noise level at low frequency (dB re 1 μPa)
Type-094 (Jin)	SSBN	140
Type-093 (Shang)	SSN	145
Type-092 (Xia)	SSBN	160
Type-091 (Han)	SSN	155

Source: (Riqiang, 2011).

Wu Riqiang. Prof. Wu notes that China undertook a combination of bastion and coastal patrol strategy to protect the SSBNs (Riqiang, 2011). Chinese naval strategists are confident that the SSBNs can operate safely within the regional waters and only seek to break out into the Pacific during a crisis, so that they can launch JL-2 SLBMs capable of reaching the US mainland (Zhao, 2018). China is also scouting for advanced technology to make their submarines quieter and feasible to deploy in the open ocean without the fear of detection. The Type-95 and Type-96 submarines currently under development are expected to join the service soon by 2025 to match up with the US naval power in the region.

The conventional, diesel-electric submarines constitute a major share of the Chinese Submarine Force. A US CSR-2021 Report on *'China Naval Modernization: Implications for U.S. Navy Capabilities – Background and Issues for Congress'* forecasts China's submarine force will grow from a total of 66 boats (4 SSBNs, 7 SSNs, and 55 SSKs) in 2020 to 76 boats (8 SSBNs, 13 SSNs, and 55 SSKs) in 2030 (O'Rourke, 2021b). China primarily relied on Russian technology and help in building conventional submarines. The *Ming*-class, *Song*-class, and first *Yuan*-class submarines are domestically to a large extent based on the Russian submarine design. Moreover, China also imported eight Kilo-class from Russia in the early 2000s to deploy against any possible threat from the US Navy in Taiwan Straits (Pomfret, 2002). The Russian *Kilo-c*lass submarine is not in the forefront of the submarine technology, but it is quite enough to challenge the US Navy's ASW systems (Kane, 2003). Moreover, the *Kalibr* cruise missile, a main weapon of Kilo-class submarines, provides a lethal advantage against the superior US Navy's surface fleet. Based on the Kilo-class submarine, China launched a new submarine in 2006 called the Yuan-class submarine. The Yuan-class is the only advanced conventional submarine reportedly with two Stirling air-independent propulsion (AIP) systems developed by the 711 Research Institute of the China Shipbuilding Industry Corporation (CSIC) in Shanghai (Hemmingsen, 2011). The *Yuan*-class also appears to have incorporated the quietening technology from the Russian-designed Kilo-class (Episkopos, 2020). The *Yuan* is the only submarine currently in production, as has China decided to streamline the production, maintenance, and training of the crew by reducing the number of different classes in services. The *Ming*-, *Song*-, and *Kilo*-class are primarily used for coastal defence. However, during war, these conventional submarines can be optimized for conducting intelligence, surveillance, and reconnaissance (ISR), and anti-surface warfare missions in the high seas. Christopher P. Carlson, argues that 'Yuan-class have long endurance and advanced capability to operate in the open ocean, and the ability to conduct long range missions' (Carlson, 2015). He points out that the low-frequency passive flank array just above the keel blocks signifies that it is designed to operate in deeper waters where this passive sonar can serve as the primary sensor (Carlson, 2015). The Yuan-class with improved sonar and AIP capabilities will provide a crucial advantage to China. But, there is a downside to AIP technology – it's relatively low speed while submerged, making it vulnerable in the open ocean against capable adversaries (Erickson, Martinson, & Dutton, 2014). Dr. Whiteman, senior editor of *Undersea Warfare Magazine*, says, 'If their distinctive characteristics are exploited by skillful operators, AIP submarines can be used to telling effect for both short-and-medium-range missions' (Whitman, 2001). The introduction of long-range anti-ship cruise missiles (ASCMs) provides conventional submarines a stand-off capability against the powerful naval force.

China-launched ASCM underwent massive upgradation with YJ-18/B, which has a range of 220–540 km. The US naval experts claim that Type-039A/B Yuan-class submarine is especially designed for carrying the ASCMs and would significantly boost China's strike capability against other navies and expeditionary capability of the PLA Navy (Gormley, Erickson, & Yuan, 2014). According to the US Department of Defence (DOD) assessment, China is expected to produce 25 or more Yuan-class submarines by 2025 (Congress, 2021). Most of the Yuan-class submarines, including the latest generation, Type-95 and Type-96, will also be equipped with YJ-18/B anti-ship cruise missiles.

Geography is a major obstacle for China in carrying out submarine operations in the open seas. To reach the Pacific Ocean, the PLA Navy has two main routes: the Miyako Strait and the Bashi channel. These two waterways are the entrance to the chain of major archipelagos enclosing the coastline beginning from the Kuril Islands, in the north, to the Philippines and

Table 2.3 PLA Navy's Submarine Launch Anti-Ship Cruise Missile

Type	Manufacturers	Range (KM)	Payload (KG)	Speed	Guidance (inertial/ terminal)
YJ-82 Series	CASIC Third Academy	42	165	Subsonic	Inertial/ active terminal guidance
3M-54 Kalibr (Club-S)	Novator (Russia)	200	200	Supersonic	INS/active
YJ-18/B	CASIC Third Academy	220–540	140–300	Supersonic	Inertial

Source: Table is derived from (Gormley, Erickson, & Yuan, 2014), (Project, 2017), (Project, 2021).

Borneo, in the south. It is complex for China to operate the submarines in the deep water of the Philippine Sea and the broader Pacific Ocean without crossing one of these chokepoints. China faces a similar geographical issue in when entering the Indian Ocean – it has to cross crucial chokepoints like Malacca, Lombok and Sunda straits. The other major challenge in the East China Sea and Yellow Sea are the active presence of the US and Japanese military who are closely monitoring the passage that passes out into the western Pacific. Japan Maritime Self-Defence Force (JMSDF) has a decade of experience in monitoring USSR submarines in the Soya Strait between Hokkaido and Sakhalin, the Tsugaru Strait between Honshu and Hokkaido, and the Tsushima Strait between Kyushu and South Korea (The Japan Times, 2015). Japan has mastered the art of operating submarines in highly topographically complicated sea floors. Today, JMSDF has the capability and training to trace any Chinese submarine movement in the region. This is one of the reasons behind China taking control over the reefs in South China Sea, which is becoming strategically important for the PLA Navy to operate submarines in the region. China has also installed an array of sensors, antennas, satellite communication, and tracking stations in the Spratly Islands to monitor the foreign navies' underwater activity in the South China Sea. This has forced the Chinese to make the Southern Theater Command as an optimal place to deploy their nuclear submarines.

The nuclear submarine base in Hainan Island is built with a special purpose to conceal submarine operations. The Type-094 Jin-class ballistic nuclear submarine and Type-093 Shang-class nuclear attack submarines are stationed closer to deep water channels that lead in and out of the South China Sea. This area is also connected to one of the busiest shipping routes connecting the Pacific Ocean with the Indian Ocean and, hence, is more suitable to conceal submarine operations. The submarine base in Hainan is now under direct control of the Central Military Commission, which is the top military decision-making body, chaired by President Xi Jinping. In 2017, President Xi appointed a veteran submariner, Vice Admiral Yuan Yubai, head of Southern Theatre Command. This was the first time a naval officer had been appointed to command the regional combat headquarters, which is responsible for the South China Sea. The appointment of admiral to the top leadership post clearly indicates the importance of South China Sea and the future roles of nuclear and conventional submarines in the Indian Ocean Region. Therefore, for safe submarine operations, it is pivotal for China to control and monitor coastal seas and secure the critical chokepoints in the Pacific Ocean and the Indian Ocean for military operations beyond the first island chain. Both passages, indeed, create opportunities to deploy submarines for patrol, to block the passage and to monitor the transit of fleets. In case of war, PLA Navy would deploy additional asserts to target and eliminate the hostile force's surface and underwater threats to Chinese ports. The Chinese submarines are crucial not only in playing an important role in denying underwater spaces for the enemy's clandestine operations but also to guaranteeing the safety of surface naval operations and merchants' vessels. China's emerging economic and trade cooperation with the littorals provide some kind of leverage to Beijing in building naval base facilities in the region. The deployment of PLA Navy's flotilla, submarines, and hydrographic vessels in the water suggest Beijing's strategic engagement with the region.

Chinese Basing Facilities, Submarine and Scientific Marine Vessels Deployment in the Indian Ocean

China envisages a stronger role for the PLA Navy in the Indian Ocean. China's growing investments in the infrastructure, port development, and economic investments in the energy and mining sector (especially in the African countries) are a reflection of China's major economic and political interest in the region (see Table 2.1). The strategic location of these ports near international sea lines of communication (SLOCs) proves that not all Chinese projects are driven by commercial logic. There is plausibility in the argument that China could well use these facilities for military purposes.

Table 2.4 Chinese Investment in Port Development in the Indian Ocean

Port (country)	Chinese companies	Year of investment
Kyaukpyu (Myanmar)	China International Trust Investment Cooperation Group	2014
Hambantota (Sri Lanka)	China Merchants Port Holdings Co. Ltd.	2008
Port Colombo (Sri Lanka)	China Harbour Engineering Company Ltd.	2014
Chittagong (Bangladesh)	China Communications Construction Company Ltd.	2015
Gwadar (Pakistan)	China Overseas Port Holding	2003
Ras Al-Khair Port Project (Saudi Arabia)	China Harbour Engineering Company Ltd.	2007
Khalifa Port (Abu Dhabi, the UAE)	COSCO Shipping Ports	2016
Industrial Zone of Duqm (Oman)	China-Arab Wanfang Co. Ltd. (Ningxia)	2016
Mombasa Port (Kenya)	China Road & Bridge Corporation	2013
Bagamoyo (Tanzania)	China Merchants Port Holdings Co. Ltd.	2013
Port Djibouti (Djibouti)	China Merchants Port Holdings Co. Ltd.	2014

Source: Compiled by author from (Sun & Zoubir, 2017), (Faridi, 2021).

The naval base in Djibouti is a cornerstone for the PLA Navy to continue its operations in the Indian Ocean. According to experts, China has built 1,120-foot piers to accommodate new aircraft carriers, or larger ships, including nuclear attack submarines if required (Sutton, 2020). Chinese submarines also visit Karachi and Colombo ports as part of their deployment in the Indian Ocean. The port development provides certain leverage to the PLA Navy to engage in maritime cooperation with these countries. The port facilities are used by China to monitor military ships, merchant vessels, and other vessels in the region.

The PLA Navy's presence in the Indian Ocean has been growing steadily since 2009, when piracy and hijacking ships for ransoms in the Gulf of Aden disturbed global energy and trade routes. A major justification for their presence is for the security of their commercial engagements and maritime trade. Since 2013, Chinese submarines have been in the Indian Ocean, and they claim to have been deployed for anti-piracy duties. Chinese submarines use Straits of Malacca, Lombok, or Sunda to reach the Indian Ocean. The Sunda strait is very shallow with a mean depth of 50 m; furthermore, the sandbanks, the presence of oil platforms in the Sunda shelf,

Table 2.5 PLA Navy's Presence in the Indian Ocean

Country	High level PLA contacts	Bilateral exercises	Port calls
Indonesia	-	✓	✓
Singapore	-	✓	✓
Thailand	-	✓	✓
Myanmar	-	✓	✓
Bangladesh	✓	✓	✓
Sri Lanka	✓	✓	✓
Maldives	✓	✓	✓
Pakistan	✓	✓	✓
Iran	✓	✓	✓
Oman	✓	✓	✓
Yemen	✓	-	✓
Djibouti	✓	✓	✓
Kenya	✓	✓	✓
Tanzania	✓	✓	✓
Mozambique	✓	-	✓
South Africa	✓	✓	✓
Madagascar	-	-	✓
Seychelles	✓	-	✓
Mauritius	-	-	✓

Source: Compiled by Author from (Congress, 2020). In addition to satellite tracking stations in Pakistan and Kenya.

and the active presence of fishing vessels render these waters difficult to navigate. The Lombok strait is deep enough for submarines to proceed in normal mode of navigation. In the case of Malacca strait, for navigation safety, Chinese submarines have to navigate on the surface. The submarines are often accompanied by the submarine tender ships that are spotted in the Colombo ports.

The deployment of submarines along with the PLA Navy's survey and hydrographic ships in the Indian Ocean has become a regular feature since 2017. According to an Indian news report, in 2017, PLA Navy's top naval survey ship – the Type-636A hydrographic survey ship Haiyang-class 22 – was surveying waters of the Indian Ocean and was in all probability charting for better submarine operations. In 2018, the PLA Navy's modern Type-625C Shiyan 3 Oceanographic Survey Ship was surveying the Makran trench in the Persian Gulf as part of its maritime exercise with the Pakistan Navy. In the same year, Xiang Yang Hong 10 was deployed in the Indian Ocean to survey China Ocean Mineral Resources R & D Association's (COMRA) contract zone in southwest Indian Ocean. Since 2019, Chinese survey ship, Xiang Yang Hong 03, has been surveying the deep waters of the Bay of Bengal, Arabian Sea, and the waters west of Indonesia, which is considered to be an important area of submarine operations for both India and Australia (Sutton, Chinese Survey Ship Caught 'Running Dark' Give Clues to Underwater Drone Operations, 2021a). The increasing presence of Chinese State-run hydrographic ships and survey vessels would facilitate future Chinese SSBNs' and SSNs' deployment in the Indian Ocean. Unlike the Pacific west, which is sprawling with the US and JMSDF submarines and ASW ships monitoring the PLA Navy's submarine movements, the Indian Ocean is relatively safe for the Chinese submarine to operate. However, bypassing choke points would likely give away Chinese submarines while entering the Indian Ocean, which will remain a major challenge for China in the future as well.

China's rush to secure its strategic resources, which are spread across the world – from the African Coast to the South American Coast and from Australia to the far east of Russia, now expanding to Arctic region – justifies not only PLA Navy's deployment in the Indian Ocean Region (IOR) but also its desire to gain foothold in the Indian Ocean. Prof. You Ji, expert on Chinese affairs, explains that, in the absence of a strong flotilla to protect the SLOCs in the Indo-Pacific region, submarine warfare is the best way to protect the SLOCs. Submarine, as a kind of contingency capability, is well tuned with the Chinese strategy to protect its shipping line in the Indian Ocean (Kerr, Harris, & Qin, 2008). Prof. Ji also argues that, unlike nuclear submarines, conventional submarines lack the efficacy to fight submarine warfare in the Indian Ocean. Moreover, building nuclear and conventional submarines is more cost-effective than building an escort fleet for the same purpose. Even though his logic stems from the understanding that the Chinese PLA Navy presence in the Indian Ocean is limited in comparison with other navies, given the geopolitical constraints China is facing, the deployment of submarines for the protection of SLOCs is the most viable strategy to deter enemy naval fleets from disrupting Chinese SLOCs in the Indian Ocean.

2. Pakistan's Sea-Based Defence Capability

Towards the turn of the twenty-first century, Pakistan's naval posture changed in response to India's growing maritime influence in the region. India's continuous development of its nuclear triad and modernization of its naval force to increase the option of survivability has raised serious concerns in Pakistan. This brought Pakistan to develop the submarine-launched cruise missile (SLCM) – as a part of Pakistan's nuclear policy. The recent development in Pakistan's Navy suggests that it is priming to build capability to launch tactical nuclear missiles from conventional submarines. The two consecutive flight tests of Babur-3 – a nuclear-capable submarine-launched cruise missile –is a clear indication that Islamabad may have developed or be in the process of achieving the sea-based nuclear strike capability. According to Pakistan's Inter-Services Publication Relations (ISPR) statement, the missile had fired to a range of 450 km from an underwater dynamic platform, and had 'successfully engaged its target with precise accuracy, meeting all the flight parameters' (ISPR, 2018). Notably, Babur-3 is said to have the capability of delivering various types of payload with advanced features like underwater controlled propulsion system, advance guidance, and navigation features, and terrain hugging technologies (ISPR, 2018). Pakistan sees the development of Babur-3, crucial to reinforce its minimum credible deterrence. Speaking at Carnegie International Nuclear Policy conference in 2015, Gen. (retd.) Khalid Kidwai, former head of Pakistan's Strategic Plans Division, acknowledged that Pakistan's second-strike capability comes from sea-based platforms (Kidwai, 2015). He also confirmed that Pakistan is making progress in building platforms and communication aspects to achieve the sea-based deterrence capability – nuclear submarine project called KPC-3 – to design and manufacture a miniaturized nuclear power plant for a submarine (Lobner, 2018).

2.1 *Status of Pakistan's Submarines*

The Pakistan maritime nuclear force is now equipped with multiple French origins, Agosta 90B, diesel-electric submarines capable of launching SLCM through torpedoes tubes. Pakistan is also upgrading Agosta-class submarines with the help of Turkish company STM to undertake a mid-life refit of one submarine with an option for the other two (STM, n.d.). The scope of modernization includes installation of SERO 250 S search periscope, OMS 200 optronic mast, Kelvin Hughes SharpEye navigation radar, Aselsan ARES-2SC ESM system, and MSH-01 sonar hydrophones from Meteksan Defence Industries (Scott, 2016). Pakistan is also in the process of acquiring Type-041 Yuan-class submarines with AIP systems from China. In 2016, Pakistan inked an agreement with China for acquiring eight Type-041 Yuan-class submarines, of which four submarines will be delivered by China Shipbuilding Trading Company (CSTC) in 2023, and the remaining four will be built in Karachi Shipyard and Engineering Works with the help of Chinese assistance. The Type-041 Yuan-class submarine with AIP system is stealthier and provides a credible underwater capability to challenge India's dominance in the northern Indian Ocean (Carlson, 2015). Tactically, the Yuan-class submarine is best suited to maintain Pakistan's maritime force because it has a unique advantage in the Arabian Sea, where marine traffic originating from the Persian Gulf can easily camouflage the acoustic signature of the submarine. The advantage of Agosta-class submarines and the induction of newer-class submarines built in China will enhance Pakistan's underwater capability against superior Indian naval force in the Arabian Sea. Moreover, the building of sea-based nuclear strike capability would provide Pakistan a perception of not falling behind India in the nuclear sphere (Calvo, 2016).

Pakistan's introduction of a sea-based nuclear force is a major destabilizing factor in the region. Christopher Clary and Ankit Panda claim that 'Pakistan's pursuit of a sea-based nuclear force introduces considerable new nuclear risk in South Asia' (Clary & Panda, 2017). The main concern for India arises from the China-Pakistan nexus, which has a history of running clandestine nuclear programmes. India also fears that the Chinese might be helping Pakistan in the miniaturization of nuclear warheads to fit atop of the Babur-3. The Chinese eagerness to help Pakistan in achieving sea-based nuclear strike capability might be part of the Chinese Primer Xi Jinping's strategy to engage with Pakistan.

3. Proliferation of Conventional Submarines

The changing geostrategic situation in the Indo-Pacific region is forcing many countries to procure submarines. Many nations including Australia, China, Indonesia, Singapore, and Iran are increasing the size of their submarine force. Malaysia, Vietnam, and Philippines – which had not previously possessed ocean-going submarines – are now acquiring them to defend against external threats. The submarine size of major navies like the

Table 2.6 Pakistan's Present and Future Underwater Capability

Class of submarine	Builder	Torpedo	Major armaments	Diesel engine	AIP technology	Combat systems type
Pakistani designated Khalid-class (Agosta Submarine)	French DCNS	04 x 533 mm torpedo tubes	SM39 Exocet (Launched from torpedo Tubes) & DM2A4 torpedo	396 SE84 series (MTU Friedrichshafen GmbH of Friedrichshafen, Germany.)	MEMSA (Module d'Energie Sous-Marin Autonome) AIP System.	Thomson-Sintra SUBTICS Mk 2.
Chinese designation Yuan-class (Pakistani designation not known)	China State Shipbuilding Industrial Corp (CSIC)	06 x 533 mm torpedo tubes	YJ-8X (C-80X) & Yu-4 (SAET-50) and Yu-3 (SET-65E)	German MTU 16V396SE84	AIP (probably fitted with Kockums Stirling AIP technology)	Unknown

Source: Compiled by author from Janes Fighting Ships (Submarines, 2021).

Table 2.7 Submarines in the Indian Ocean Region

Country	Number
Oceania	
Australia	
SSK – Collins	6
South China Sea	
Indonesia	
SSK –	
Cakra (Type-209/1300)	2
Nagapasa (Type-209/1400)	2
Malaysia	
SSK	
Tunku Abdul Rahman (FRA Scorpène)	2
Singapore	
SSK	
Challenger (ex-SWE Sjoormen)	2
Archer (ex-SWE Västergötland) (AIP fitted)	2
Vietnam	
SSK	
Hanoi (RUS Varshavyanka)	6
SSI	2
Yugo (DPRK)	
Bay of Bengal	
Bangladesh	
SSK – Nabajatra (ex-PRC Ming Type-035G)	2
Myanmar	
SSK – UMS MinyeTheinkhathu	1
Northern Indian Ocean/Arabian Sea	
Iran	
SSK – Taregh (RUS Paltus Project 877EKM)	1
SSC – Fateh	1
SSW	14
Ghadir	1
Nahang	

Source: Table is based on IISS Military Balance 2021 (Asia, 2021).

United States, China, Japan, and India is also increasing and has been regularly deployed for patrol in the Indian Ocean. According to the UN Arms Register of Conventional Arms, there is no limit on the sale of conventional submarines, but nations are required to declare the export of submarines with a standard displacement of 500 metric tons or above, and equipped for launching missiles with a range of at least 25 km or torpedoes with similar range (The Global Reported Arms Trade: Transparency in Armaments through the United Nations Register of Conventional Arms, 2017).

In 2003, states in the Wassenaar Arrangements further reduced the threshold for reporting submarines export to 150 metric tons or above, equipped with missiles or torpedoes with a range of 25 km or more (Wassenaar Arrangement on Export Controls for Conventional Arms and Dual-Use Goods and Technologies, Public Documents Volume I, 2019). While 500 metric tons cover most of the submarine export, non-proliferation experts claim that the category listed under warship is 'vague' and 'amorphous', making the reporting quite inaccurate (Moltz, 2005). Furthermore, the gaps in the international export control regime regarding submarines being exploited by the major weapon-exporting nation pose risk to regional security. Therefore, it is difficult to account for the exact sale of submarines in this region.

The small and medium powers in the Indian Ocean Region are steadily modernizing various aspects of its naval structure to keep the sea lane open, conduct naval intelligence, and reconnaissance in the waterway to keep the passage clear from disruption. Most importantly, the submarines can provide 'asymmetric capability' against the most powerful naval adversaries. Taking into account China's strategic ramification, Australia's Defence White Paper 2016 laid a framework to modernize its maritime force. At present, the Royal Australian Navy operates six *Collins*-class diesel-electric submarines specifically designed for the defence and surveillance of the Indian Ocean as well as the Pacific Ocean (The Collins Class, n.d.). In view of the emerging challenge in the Indo-Pacific region, on September 15, 2021, Australia, the United States, and UK signed a trilateral security pact and decided to extend support to uplift the underwater capability of Australia through extending support to share nuclear-powered propulsion system with Canberra. Under this pact, Australia has planned to acquire at least eight nuclear-powered submarines, leveraging decades of experience from the United States and UK (Media Statement, 2021). This led to the cancellation of an earlier deal with the French DCNS to acquire 12 new *Attack-class* submarines to replace Collins-class diesel-electric submarines. The shift from the conventional Attack-class submarine to nuclear-powered submarine clearly signifies the growing cooperation on the security and defence capabilities, which will enhance joint capability and interoperability. This would further cause the Royal Australian Navy to share data with American submarines and other US assets during combined operations in the future.

The littorals of South China Sea are also upgrading their maritime capabilities, particularly undersea warfare systems for the surveillance and reconnaissance of their EEZ, which is constantly under threat from the Chinese Navy. The Indonesian Navy – also known as *Tentar Nasional Indonesia-Angkatan Laut* (TNI-AL) – once operated 12 Whiskey-class submarines purchased from the Soviet Union in the 1970s but now only operates 2 classes of submarines: the older Cakra-class Type-209/1300 vessels and the Nagapasa-class Type-209/1400 vessels. As part of Indonesia's Defense Strategic Plan 2024, Jakarta had signed a deal with Daewoo Shipbuilding and Marine Engineering (DSME) in 2019 to procure three more Nagapasa-class submarines (Gady, 2019). In April 2021, Indonesia lost German-built KRI-Nanggala 402 off the coast of Bali during a torpedo exercise, raising serious concerns about the modernization plan of TNI-AL. Singapore, a city state, also operates four diesel submarines and is about to acquire four Type-218 SG submarines from Germany. The first of the four Type-218 S-class submarine, RSS *Invincible*, was launched in Feb 2019, and is currently undergoing sea trials. The Type-218 SG submarines are slated to replace Challenger- and Archer-class submarines that the RSN has operated for more than two decades (Zhang, 2019). The Type-218 SG submarines are designed for Singapore's operating environment, particularly for the shallow waters and busy sea routes. All the four Type-218 S-class submarines are fitted with AIP and would provide the Singapore Navy a strategic capability against emerging maritime threats in the region. The Vietnam Navy operated SSK 6 Hanoi (RUS Varshavyanka) is a Russian-imported Kilo-class submarine with *Klub-S* (3M-54) anti-ship cruise missiles and two Yugo midget-class submarines – imported from DPRK – and represents a potential major subsurface challenge to the Chinese naval force in the region. The Malaysian Navy acquired two submarines *Tunku Abdul Rahman* (FRA Scorpène) from France in the early 2000s. The basing of both submarines at Kota Kinabalu Naval Base in Sabah, East Malaysia, indicates that the primary role is to protect Malaysian maritime interest in the South China Sea (Beng, 2014).

In the Bay of Bengal, submarines are gaining prominence. India has an experience of operating submarines for more than four decades, while Bangladesh and Myanmar have recently acquired submarine capabilities. Prominent maritime expert Vijay Shakhuja says that the 'mutual suspicions' among the navies in the Bay of Bengal has led Bangladesh, Myanmar, and Thailand to acquire submarine capability (Sakhuja, 2020). He also mentioned that the submarines constitute changes to the overall Order of Battle (ORBAT) among the Bay of Bengal navies. In 2017, Thailand Cabinet approved the purchase of its first submarine from China for 434.1 million US dollars (Strangio, 2020). But, following the Covid pandemic and shrinking of Thailand's economy, the government has put the submarine deal on hold. Myanmar inducted a 3,000 tons diesel-electric *Kilo*-class Submarine UMS Minye Theinkhathu, which was gifted by India to gain edge over China in the Bay of Bengal (Laskar, 2020). Bangladesh has acquired two Type-035 G-class submarines also known as *Ming-c*lass from China. So far, Bangladesh has not encountered any major maritime threat in the region, the purchase of submarines from China only indicates the growing strategic partnership between the two counties (Raghuvanshi, 2017). Moreover, the proliferation of submarines in the Bay of Bengal is a major concern for India, which has a significant naval presence in the region.

In the northern Indian Ocean, other than Pakistan, Iran has acquired three Russian *Kilo*-class submarines called Taregh-class in Iran. Iran also launched an indigenously built submarine called *Fateh*-class diesel-electric submarine in 2017. This submarine can dive up to 200 m and has an endurance of 5 weeks (Binnie, 2019). The Fateh-class submarine is equipped with Nasr-1 (anti-ship cruise missile), which is based on the Chinese C-704, which has a range of 30 km. Admiral John Miller, a retired US Navy vice admiral, says that 'The successful production and employment of an Iranian produced anti-ship cruise missile from an Iranian-produced submarine is a significant event that potentially challenges the stability of the entire Gulf region' (Miller, 2019). Iran is also operating several midget-class submarines, such as the Ghadir-class based on the DPRK design. The exact numbers, the armaments, and offensive capability of the Ghadir-class is still unclear (Nadimi, 2020). The sources from western media only point out the asymmetric capabilities of these midget submarines, which pose a serious threat to regional shipping as well as to the US aircraft carriers in the region.

The growing submarine fleet of different nations raises the risk of underwater incidents/accidents in the Indian Ocean. Many Southeast Asian navies use the submarine for collecting intelligence and reconnaissance missions in disputed water, which raises the risk of conflict between nations. Naval expert, Sam Bateman, explains the danger associated with operating submarines in such close quarters (Bateman, 2011). First, submarines are dangerous systems, even a small accident onboard might have catastrophic consequences. Second, the increased number of submarines operating in the region pose a navigational risk. Particularly, submarine activity near busy sea lanes or fishing-activity zones might endanger submarine operations. If multiple submarines are operating in one area, it would require highly skilled manpower to avoid accidents between the submarines. Third, submarines deployed in clandestine operations in the disputed water heightens the risk of detection by another country's anti-submarine systems, which might result in severe political ramifications. India on many occasions has raised objections to foreign navies conducting oceanographic activity for submarine operation in its EEZ. The proliferation of conventional submarines with AIP systems onboard makes detection highly impossible. Given the peculiarity of the geophysical condition of the Indian Ocean, it will be difficult to detect conventional submarines in the region, posing a serious security and safety concern for India.

4. Unmanned Underwater Vehicles in the Indian Ocean

The unmanned/autonomous systems are gaining prominence among the strategic thinkers and military planners. In the maritime domain, the UUVs have potential to change the underwater warfare with the entry of various new technologies. The UUVs come in different sizes, from small to large. UUVs offer flexibility to carry various payloads from sensors to weapon systems because they were designed not to incorporate space for equipment, support facilities for onboard human operators. The operating cost of a UUV is also relatively less expensive than many frontline naval ships; new research on UUV shows that it can be deployed in long missions or perform dangerous missions with absolutely no casualties. The United States and China are making steady progress in UUV technology and robotics.

The US Navy's Unmanned Maritime Systems classifies the UUV programme into four categories: small, medium, large, and extra large. The US Navy has requested 580 million US dollars in the FY2021 for the research and development of larger UUVs and its associated technologies (O'Rourke, Navy Large Unmanned Surface and Undersea Vehicles: Background and Issues for Congress, 2021a). The US Navy has currently deployed ASW Continuous Trail Unmanned Vessel (ACTUV), otherwise popularly known as 'Sea Hunter,' after extensive sea trials. The ACTUV modular design allows for the integration of flexible mission payloads suitable for a range of tasks. The navy is also planning to acquire medium to large autonomous vehicles as part of its fleet architecture. This means that the US Navy might operate medium or large autonomous vehicles or UUVs in the Indian Ocean as part of its deployments.

China is also making significant progress in underwater drone technology. The Chinese government and PLA are funding 15 universities and technical institutions in China to conduct research on UUVs/unmanned surface vessels (USVs) (Chase, Gunness, Morris, Berkowitz, & Purser, 2015). Chinese scholars also see the greater importance of UUVs/USVs in territorial disputes. China's hydrographic and survey ships have been using underwater drones/gliders as part of its maritime exploration in the Indian Ocean. In December 2020, an Indonesian fisherman picked up sea wing gliders belonging to China in waters near the Selayar Islands, in the South Sulawesi province, which is on the eastern side of the Makassar Strait, close to the international Sea Line of Communication (SLOC) (Kannan, 2021). H I Suttons – a writer, illustrator, and analyst who specializes in submarines and subsurface systems – in his analysis published in the United States Naval Institute (USNI) News, indicates that the Chinese have been operating underwater drone 'Sea Wing' in the Indian Ocean (Sutton, Two Chinese

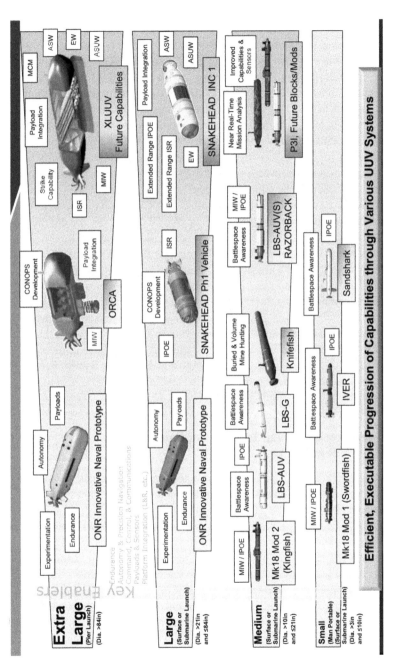

Figure 2.1 The US Navy: Unmanned Maritime Systems Update

Source: (Small, 2019).

Survey Ships are Probing a Strategic Section of the Indian Ocean, 2021b). China is also looking to develop the US Navy's 'Sea Hunter' equivalent, an ASW autonomous vehicle for long missions. The PLA Navy is also fielding the UUV system HSU 001. HSU 001 was revealed during the 70th anniversary of the founding of the People's Republic of China, on October 1, 2019. HSU 001 is approximately 5 m in length and 1.5 m in width, with twin propellers, which can maximize the range. The twin mast, which can fold back into the hull, is believed to be an 'advanced electro-optical detection system' (Goldstein, 2020). The system is well suited for ISR capability, has long endurance, and the capability to provide round-the-clock watch in the South China Sea. Lyle J. Goldstein, research professor in the China Maritime Studies Institute (CMSI) at the United States Naval War College, observes that Chinese naval strategists might cultivate undersea AI and highly capable UUVs to 'circumvent their long-recognized weakness in submarine warfare' (Goldstein, 2020).

Additionally, UUVs and USVs could augment the PLA's capabilities in several areas, such as the use of space assets and other ISR capabilities for targeting over long distances. The advancements in battery and fuel cell technology would enable the endurance of UUVs and can operate deep in the ocean, thereby changing the nature of ASW. The US and NATO forces are developing the next level of autonomous underwater drones with AI features to detect, track, and identify the quietest enemy submarine even in harsh sea conditions. The emerging ASW technologies appropriate to tackle threat scenarios need to be identified and taken up for development, procurement, or co-development.

5. Threat Accidents and Incidents of Submarines in the Indian Ocean

The deployments of nuclear submarines in the Indian Ocean also aggravate the risk of accidents/incidents in the region. India fears that Chinese SSBNs and SSNs operating in the Indian Ocean may escalate the risk of nuclear incident/accidents in the region. Accidents involving nuclear submarines may cause a major underwater challenge to India's maritime security. The nuclear submarines are designed and built to operate in highly hostile conditions. The modern-day nuclear submarine's design and the crew's training in advanced safety features will help to control these damages even if an accident occurs. The criticality of these accidents – including uncontrolled fire, explosion, and severe radiation or sinking of ships due to collisions – may lead to nuclear catastrophes at sea. The advancements in nuclear technology and submarine designs have improved the safety features of modern nuclear submarines. A majority of incidents/accidents of nuclear submarines occur close to the submarine base or on the international straits. The international straits like Gibraltar, Hormuz, Tsushima, and Juan de Fuca are close to submarine bases, are open to huge fishing traffic, and are hotspots for submarine-related accidents (Alexander, 1992). According to the report prepared for the Outrider Foundation by the University of Wisconsin-Madison, the consequences of nuclear submarine accidents are fairly local and in no way will approach the magnitude of the Chernobyl incident (Benda, Chandra, Janoska-Bedi, Kane, & Vannucci, 2019).

However, history has shown that accidents/incidents at nuclear submarines are unavoidable. In July 2019, a Russian nuclear submarine called Losharik AS-12 was involved in an accident that killed 14 Russian sailors and officers aboard. Russia was successfully able to recover the damaged submarine, Losharik AS-12. Since World War II, the Soviet Union or Russia and the United States have lost 9 nuclear-powered submarines to the bottom of the sea (Rychlak, 2019). So far, the Indian Ocean has not witnessed any kind of nuclear submarine incident/accident in recent times. The last reported nuclear accident was in 1977 when the Soviet Echo II-class cruise missile nuclear submarine suffered a fire in the Indian Ocean. The submarine's fire took several days to extinguish. It was finally towed away by the Soviet Trawler to the Vladivostok naval base. This incident did not cause any nuclear exposure, but it is important to understand – from the context of growing Chinese SSBN and SSN deployment in the Indian Ocean – that it may heighten the risk of nuclear accidents/incidents at high seas. Other than nuclear submarines, conventional-diesel submarine accidents in the Indian Ocean are another major concern as nearly ten countries in the region have acquired submarines. Indonesia's KRI-Nanggala 402 accident in April 2021 and USS Connecticut accident in the South China Sea in October 2021 are grim reminders of the future underwater challenges this region is likely to face. Moreover, the future deployments of UUVs/autonomous vehicles of the United States, China, and other NATO countries are most likely to increase the risk of accidents in the region.

In conclusion, the emerging underwater challenges in the Indian Ocean are of serious concern to India's maritime security. The Chinese naval ambition to expand its presence both in the Pacific Ocean and in the Indian Ocean would significantly escalate tension between the United States, China, Japan, and other democratic nations. Similarly, Pakistan's plan to acquire newer submarines is of great concern for India in terms of protecting its maritime boundaries. The upcoming Jinnah Naval Base in Makran Coast, which is likely to become the future submarine base for Pakistan located close to the international SLOCs, indicates Pakistan Navy's growing interest in the underwater domain. The many littorals of Indo-Pacific countries have acquired submarines to enhance their maritime domain awareness in the region. This raises the risk of safety and security of submarine operations in the region. Therefore, it is important for India to enhance its underwater surveillance capability through building robust ASW capabilities and to mitigate other subsurface threats in India. To develop proper strategies, it is important that India understand the evolution of undersea warfare and the surveillance capabilities of the major powers for effective detection and track of submarines in the deep oceans.

Note

1 Bell (2009) indicates that Type-094 design factors, including shape, skin friction, flood openings, and propellers, are major sources of noise.

Bibliography

Alexander, L. M. (1992). The Role of Choke Points in the Ocean Context. *GeoJournal*, 503–509.

Asia. (2021). *The Military Balance* (pp. 218–313). London: Routledge.

Bateman, S. (2011). Perils of the Deep: The Dangers of Submarine Proliferation in the Seas of East Asia. *Asian Security*, 61–84.

Bell, S. D. (2009). *The Impact of the Type 094 Ballistic Missile Submarine on China's Nuclear Policy*. Naval Postgraduate School, the US Navy. Retrieved from: https://calhoun.nps.edu/handle/10945/4700.

Benda, M., Chandra, A., Janoska-Bedi, S., Kane, J., & Vannucci, J. (2019). *Silent Dangers Assessing the Threat of Nuclear Submarines*. Madison: University of Wisconsin.

Beng, A. (2014). Submarine Procurement in Southeast Asia: Potential for Conflict and Prospects for Cooperation. *Pointer – Journal of Singapore Armed Forces*, 55–66.

Binnie, J. (2019, February 18). Iran Commissions Fateh Submarine. *Jane's 360*.

Calvo, A. (2016, March 23). *Pakistan's Navy: A Quick Look*. Center for International Maritime Security. Retrieved from: https://cimsec.org/pakistans-navy-quick-look/ (Accessed on April 12, 2021).

Carlson, C. P. (2015, August 31). Inside the Design of China's Yuan-Class Submarine. *USNI News*. Retrieved from: https://news.usni.org/2015/08/31/essay-inside-the-design-of-chinas-yuan-class-submarine (Accessed on December 11, 2021).

Chan, M. (2021, May 2). China's New Nuclear Submarine Missiles Expand Range in US: Analysts. *South China Morning Post*. Retrieved from: www.scmp.com/news/china/military/article/3131873/chinas-new-nuclear-submarine-missiles-expand-range-us-analysts (Accessed on August 11, 2021).

Chase, M. S., Gunness, K., Morris, L. J., Berkowitz, S. K., & Purser, B. (2015). *Emerging Trends in China's Development of Unmanned Systems*. Santa Monica: RAND Corporation.

Clary, C., & Panda, A. (2017). Safer at Sea? Pakistan's Sea-Based Deterrent and Nuclear Weapons Security. *The Washington Quarterly*, 149–168.

The Collins Class. (n.d.). *Submarine Institute of Australia*. Retrieved from: www.submarineinstitute.com/submarines-in-australia/The-Collins-Class.html (Accessed on November 1, 2021).

Congress, A. R. (2021). *Military and Security Developments Involving the People's Republic of China 2020*. Washington: Office of the Secretary of Defence.

Congress, R. T. (2020). *U.S.-China Economic and Security Review Commission*. Washington: U.S.-China Economic and Security Review Commission.

Episkopos, M. (2020, November 19). China's AIP Stealthy Submarine Force: A Worry for the U.S. Navy? *The National Interest*. Retrieved from: https://nationalinterest.org/blog/buzz/chinas-aip-stealthy-submarine-force-worry-us-navy-172869 (Accessed on February 4, 2021).

Erickson, A. S., Martinson, R., & Dutton, P. (2014). *China Near Seas Combat Capabilities*. Newport, RI: China Maritime Studies Institute, U.S. Naval War College.

Faridi, S. (2021, August 19). China's Ports in the Indian Ocean. *Gateway House*. Retrieved from: www.gatewayhouse.in/chinas-ports-in-the-indian-ocean-region/ (Accessed on September 11, 2021).

Gady, F. S. (2019, April 15). Indonesia, South Korea Ink $1 Billion Contract for 3 Diesel-Electric Submarines. *The Diplomat*. Retrieved from: https://thediplomat.com/2019/04/indonesia-south-korea-ink-1-billion-contract-for-3-diesel-electric-sub (Accessed on October 11, 2021).

Gokhale, N. A. (2015). *1965 Turning the Tide*. New Delhi: Bloomsbury.

Goldstein, L. J. (2020, March 28). China Hopes UUVs Will Submerge Its Undersea Warfare Problem. *The National Interest*. Retrieved from: https://nationalinterest.org/blog/buzz/china-hopes-uuvs-will-submerge-its-undersea-warfare-problem-138597 (Accessed on July 1, 2021).

Gormley, D. M., Erickson, A. S., & Yuan, J. (2014). A Potent Vector: Assessing Chinese Cruise Missile Developments. *Joint Force Quarterly*, 98–105.

Hemmingsen, T. (2011, April 20). *Enter the Dragon: Inside China's New Model Navy*. Janes Navy International. Retrieved from: http://www.andrewerickson.com/wp-content/uploads/2014/02/China-Air-Space-Based-ISR_Chinas-Near-Seas-Combat-Capabilities_CMS11_201402.pdf.

India Today. (2019, August 27). Jaish Underwater Squad's Terror Plot Revealed: How Serious Is the Threat?. *India Today*. Retrieved from: www.indiatoday.in/programme/5ive-live/video/jaish-underwater-squad-s-terror-plot-revealed-how-serious-is-the-threat-1592263-2019-08-27 (Accessed on December 11, 2020).

ISPR. (2018, March 29). Pakistan Conducted Another Successful Test Fire of Indigenously Developed Submarine Launched Cruise Missile Babur Having a Range of 450 kms. *ISPR*. Retrieved from: www.ispr.gov.pk/press-release-detail.php?id=4660 (Accessed on March 22, 2021).

The Japan Times. (2015, November 26). Japan's Crack Submarine Fleet. *The Japan Times*. Retrieved from: www.japantimes.co.jp/opinion/2015/11/26/commentary/japan-commentary/japans-crack-submarine-fleet/#.Xcu0MC2B3BI (Accessed on March 10, 2021).

Kane, T. M. (2003). Dragon or Dinosaur? Nuclear Weapons in a Modernizing China. *Parameters: Journal of the US Army War College*, 33.

Kannan, S. (2021, January 9). Deep-Sea Skullduggery: China May Be Using Its Submarine Drones for Undersea Recce in Indonesian Waters. *Indian Today*. Retrieved from: www.indiatoday.in/news-analysis/story/deep-sea-skullduggery-china-may-be-using-its-submarine-d (Accessed on June 22, 2021).

Kerr, P., Harris, S., & Qin, Y. (2008). Conclusion: Tactical or Fundamental Change? In P. Kerr, S. Harris, & Y. Qin (eds.), *China's "New" Diplomacy* (pp. 229–247). New York: Palgrave Macmillan.

Khattak, T. M. (2018, December 8). Ghazi's Eternal Patrol in 'Zone Victor'. *The News*. Retrieved from: www.thenews.com.pk/print/403133-ghazi-s-eternal-patrol-in-zone-victor (Accessed on December 6, 2020).

Kidwai, K. (2015, March 23). *Pakistan's National Command Authority* (P. Lavoy, Interviewer).

Laskar, R. H. (2020, October 21). India Gifts a Submarine to Myanmar, Gains Edge Over China. *Hindustan Times*. Retrieved from: www.hindustantimes.com/india-news/india-gifts-a-submarine-to-myanmar-gains-edge-over-china/story-fblOtZ Ry3hOaJDi6CKkjuK.html (Accessed on April 22, 2021).

Lobner, P. (2018, July). Marine Nuclear Power: 1939–2018. *Lynceans.org*. Retrieved from: https://lynceans.org/wp-content/uploads/2018/07/Marine-Nuc lear-Power-1939-2018_Part-5_China-India-Japan-Others.pdf (Accessed on April 28, 2021).

Majumdar, D. (2016, June 27). Why the US Navy Should Fear China's New 093B Nuclear Attack Submarine. *The National Interest*. Retrieved from: https://national interest.org/blog/the-buzz/why-the-us-navy-should-fear-chinas-new-093b-nu clear-attack-16741 (Accessed on December 14, 2021).

Media Statement. (2021, September 16). *Prime Minister of Australia*. Retrieved from: www.pm.gov.au/media/australia-pursue-nuclear-powered-submarines-through-new-trilateral-enhanced-security (Accessed on November 12, 2021).

Miasnikov, E. (1995). *The Future of Russia's Strategic Nuclear Forces Discussions and Arguments*. Dolgoprudny: Center For Arms Control, Energy, and Environmental Studies. Retrieved from: https://spp.fas.org/eprint/snf0322.htm (Accessed on January 13, 2021).

Miller, J. (2019, March 16). Iran's New Threat to Ships in the Gulf. *IISS*. Retrieved from: www.iiss.org/blogs/analysis/2019/03/iran-new-anti-ship-missile-test (Accessed on June 1, 2021).

Moltz, J. C. (2005, June). Serious Gaps Emerging in Export Controls on Submarines. *NIS Export Control Observer, James Martin Center for Nonproliferation Studies*, 24.

Nadimi, F. (2020, April 24). *Iran's Evolving Approach to Asymmetric Naval Warfare: Strategy and Capabilities in the Persian Gulf*. The Washington Institute. Retrieved from: www.washingtoninstitute.org/policy-analysis/irans-evolving-approach-asymmetric-naval-wa (Accessed on June 5, 2021).

Office of Naval Intelligence, U. N. (August 2009). *PLA-Navy: A Modern Navy With Chinese Characteristics*. Suitland: Office of Naval Intelligence, US Navy. Retrieved from: https://irp.fas.org/agency/oni/pla-navy.pdf (Accessed on January 11, 2021).

O'Rourke, R. (2021a). *Navy Large Unmanned Surface and Undersea Vehicles: Background and Issues for Congress.* Washington: Congressional Research Service.

O'Rourke, R. (2021b, October). *China Naval Modernization: Implications for U.S. Navy Capabilities – Background and Issues for Congress.* Congressional Research Service. Retrieved from: https://sgp.fas.org/crs/row/RL33153.pdf (Accessed on November 11, 2021).

Pomfret, J. (2002, June 25). China to Buy 8 More Russian Submarines. *The Washington Post.* Retrieved from: www.washingtonpost.com/archive/politics/2002/06/25/china-to-buy-8-more-russian-submarines/eaec2e3e-fe7a-47a4-ba29-ac42a18f60bb/ (Accessed on December 11, 2021).

Project, M. D. (2017, January 9). *3M54 Kalibr/Club (SS-N-27).* Missile Threat, Center for Strategic and International Studies. Retrieved from: https://missilethreat.csis.org/missile/ss-n-27-sizzler/ (Accessed on November 1, 2021).

Project, M. D. (2021, July 28). *YJ-18.* Missile Threat, Center for Strategic and International Studies. Retrieved from: https://missilethreat.csis.org/missile/yj-18/ (Accessed on November 1, 2021).

Raghuvanshi, V. (2017, August 8). Purchase of Chinese Subs by Bangladesh 'An Act of Provocation' Toward India. *Defense News.* Retrieved from: www.defensenews.com/naval/2016/11/23/purchase-of-chinese-subs-by-bangladesh-an-act-of-provocation-towar (Accessed on May 19, 2021).

Riqiang, W. (2011). Survivability of China's Sea-Based Nuclear Forces. *Science & Global Security,* 91–120.

Rychlak, R. J. (2019, July 11). The Russian Nuclear Submarine Accident. *The EPOCH Times.* Retrieved from: www.theepochtimes.com/author-ronald-j-rychlak (Accessed on July 29, 2021).

Sakhuja, V. (2020). Submarines Gain Precedence in Bay of Bengal Naval Order of Battle. *National Security,* 27–33.

Scott, R. (2016, October 25). *New Periscope, Optronic Mast for Pakistan Agosta 90B Submarines.* Washington: Navy International.

Singh, A. (2021, March 17). Aggressive Sea Control Isn't an Option for India's Navy. *The Interpreter.* Retrieved from: www.lowyinstitute.org/the-interpreter/aggressive-sea-control-isn-t-option-india-s-navy (Accessed on November 1, 2021).

Small, C. P. (2019, January 15). *Unmanned Maritime Systems Update.* The US Navy. Retrieved from: www.navsea.navy.mil/Portals/103/Documents/Exhibits/SNA2019/UnmannedMaritimeSys-Small.pdf?ver=%202019-01-15-165105-297 (Accessed on October 28, 2021).

STM. (n.d.). Pakistan Navy AGOSTA 90B Class Submarine Modernization Project Khalid Class Agosta PNS Hamza. *STM.* Retrieved from: www.stm.com.tr/en/our-solutions/naval-engineering/pakistan-navy-agosta-90b-class-submarine-modernization-project (Accessed on June 12, 2021).

Strangio, S. (2020, September 2). Thailand Delays Controversial Chinese Sub Purchase. *The Diplomat.* Retrieved from: https://thediplomat.com/2020/09/thailand-delays-controversial-chinese-sub-purchase/ (Accessed on August 11, 2021).

Submarines. (2021). *Janes Fighting Ships*. Retrieved from: https://www.janes.com/defence-news/naval-weapons.

Sun, D., & Zoubir, Y. (2017). Development First: China's Investment in Seaport Constructions and Operations Along the Maritime Silk Road. *Asian Journal of Middle Eastern and Islamic Studies*, 35–47.

Sutton, H. I. (2020, May 10). Satellite Images Show That Chinese Navy Is Expanding Overseas Base. *Forbes*. Retrieved from: www.forbes.com/sites/hisutton/2020/05/10/satellite-images-show-chinese-navy-is-expanding-overseas-base/?sh=521863056869 (Accessed on March 18, 2021).

Sutton, H. I. (2021a, January 16). Chinese Survey Ship Caught 'Running Dark' Give Clues to Underwater Drone Operations. *USNI News*. Retrieved from: https://news.usni.org/2021/01/16/chinese-survey-ship-caught-running-dark-give-clues-to-underwater-drone-operations (Accessed on May 19, 2021).

Sutton, H. I. (2021b, March 23). Two Chinese Survey Ships Are Probing a Strategic Section of the Indian Ocean. *USNI News*. Retrieved from: https://news.usni.org/2021/03/23/two-chinese-survey-ships-are-probing-a-strategic-section-of-the-indian-ocean (Accessed on July 1, 2021).

United Nations. (2017). *The Global Reported Arms Trade: Transparency in Armaments through the United Nations Register of Conventional Arms*. New York: United Nations.

Washington, B. (2017). The South China Sea and Nuclear Deterrence in the Asia Pacific. In *Project on Nuclear Issues* (pp. 1–136). Lanham: Rowman & Littlefield.

Wassenaar Arrangement Secretariat. (2019). *Wassenaar Arrangement on Export Controls for Conventional Arms and Dual-Use Goods and Technologies, Public Documents Volume I*. Wassenaar Arrangement Secretariat. Retrieved from: https://inecip.org/wp-content/uploads/Acuerdo-Wassenaar.pdf.

Whitman, E. (2001). AIP Technology Creates a New Undersea Threat. *Undersea Warfare*, 7–13.

Zhang, L. M. (2019, February 19). Singapore Navy Launches First of Its Four New Submarines. *The Straits Times*. Retrieved from: www.straitstimes.com/singapore/spore-navy-launches-first-of-its-four-new-submarines (Accessed on May 18, 2021).

Zhao, T. (2018). *Tides of Change: China's Nuclear Ballistic Missile Submarine and Strategic Stability*. Washington: Carnegie Endowment for International Peace.

3 The Evolution of Undersea Surveillance

Submarine warfare has been a major focal point for navies since World War I. The European nations and the United States pioneered research in submarine warfare and soon understood the importance of undersea surveillance in the early twentieth century. The US and European powers advanced their hydrophones and passive sonar technology after the end of World War I. Even though the technology was undeveloped, it offered a great sense of security against the German U-boats during World War II. During the Cold War, both the United States and Soviet Union made significant discoveries in undersea surveillance and employed effective measures to detect and track each other's submarines as well as to protect their own submarine from anti-submarines weapons. However, during the Cold War, the anti-submarine warfare (ASW) was primarily focused on detecting ballistic missile submarines (SSBNs) and nuclear-powered submarine (SSNs) in the high seas of the Atlantic and Pacific Ocean. Because, by the 1970s, the land-based Intercontinental Ballistic Missiles (ICBMs) were no longer immune or off limits to attacks, more emphasis was laid on speed, endurance, quietness, and stealth features while submarine building.

As a result, the potential gaps in undersea surveillance became evident as the United States and Soviet Union were constantly upgrading their submarine propulsion technologies and other stealth features to evade the ASW barriers and sea-based sensor systems. Moreover, the diesel submarine equipped with air-independent propulsion (AIP) technology, which is part of the Soviet Union anti-access strategy, was developed in response to disrupt the NATO Navies' sea strike mission and challenge their position in projecting power in the Barents Sea area. Since the end of the Cold War, the American strategy has shifted focus on to West Asia and China, the US naval experts' focus has also shifted to the littorals of the Indo-Pacific region. Unlike the Cold War where the submarine operations were largely focused on SSBNs/SSNs in the high seas, with the maritime conflict, the

DOI: 10.4324/9781003298380-3

submarine operations are moving close to the straits, littorals, and islands. In the post-Cold War world the functions of the submarine missions have diversified to include a range of littoral activities. With that, the undersea environment has also witnessed a drastic change in the post-Cold war period. The challenges in ASW operation in littorals can be observed from two important events. From the 1982 Falklands War and from the sinking of South Korea's Cheonan ASW Corvette by Democratic People's Republic of Korea's (DPRK's) midget submarine in 2010. The 1982 Falklands War provides an understanding to the western navies about the submarine warfare in the coastal waters – particularly, that the diesel-electric submarines can pose a major risk to advanced navies fighting the Soviet Union's nuclear submarines in the Atlantic. The Falklands War erupted between Britain and Argentina in the South Atlantic Ocean, over an island 400 miles east of Argentina. The British naval force deployed three SSNs against two Argentinian diesel submarines and clearly held the upper hand in the ASW operations. During the initial phase, Argentina lost one submarine to the British naval force. With one confirmed kill, the UK submarine still faced an enormous challenge from a single diesel submarine, *San Luis*. This is certainly true given the difficulties of sound propagation in the littorals, but the numerical advantage of three to one did nothing to aid the British fleet in curtailing the efforts of the Argentinian SSK (Pittman, 2008). In many incidents, *San Luis* came close to attacking Britain's surface, but then the submarine withdrew due to technical malfunction. If the *San Luis's* weapon had not malfunctioned, Britain could have suffered huge losses in the Falklands War.

In a similar incident, in 2010, Republic of Korea (ROK) Navy's Cheonan, ASW Corvette, armed with fairly modern hull-mounted sonar, Mk-46 mod 2 light torpedoes, and Mk-9 depth charge racks was torpedoed by DPRK's midget submarine in the Yellow Sea. In this incident, 46 crew members onboard of the *Cheonan* lost their lives. The DPRK's midget submarine is considered to be of an obsolete design by western standards, but it managed a sneak attack on the fairly advanced ASW Corvette in familiar coastal waters.

The two incidents show that modern ASW is not adequately capable of detecting smaller submarines in coastal waters. The lessons from the Falkland War and 2010 *Cheonan* incident are numerous, including the following: the threat posed by modern diesel submarines or quiet submarines with asymmetric capability has no defence due to the ineffectiveness of current sonar sensors to accurately identify submarines in shallow waters. The littoral water is crowded with many activities, posing a major challenge to submarine operations as well as to the detection of submarines. The US

Navy understood the consequence of littoral warfare, the US Navy's 1992 White Paper, '*Forward From the Sea*' signalled a change in focus as the priorities of the naval services shifted from the high seas to project power in the littoral regions of the world, subsequently the navy's spending on littoral warfare also saw significant rise (Forward from the Sea, 1992). In the following years, in the 1998 Naval Doctrine Command, the concept of 'Littoral Combat Ship Open Ocean Anti-Submarine Warfare' was released (Littoral Anti-Submarine Warfare Concept, 1998). This became a fundamental document on the undersea warfare strategy for the US Navy, because the major discussion in the concept paper was around undersea warfare, and it laid out a plan for the US Navy to counter underwater threats in the littoral waters.

Given the growing significance of undersea warfare, it is important for India to examine the evolution and advances in technology and change in ASW strategy since the Cold War years. This will help India to understand the capabilities and limitations in undersea surveillance. As discussed in Chapter 2, India is facing a unique challenge in the Indian Ocean. It has to deal with a wide range of underwater challenges ranging from monitoring adversarial country's submarine activity in the Indian Ocean, to safety and security risk posed by the Southeast Asian country's naval submarines operating close to chokepoints and international sea lanes in the region. Naval experts, security professionals, and strategists are taking note of the gaps in India's maritime security and safety and are addressing ways to reduce those gaps. Therefore, this chapter will largely discuss the United States' and Soviet Union's undersea surveillance capabilities and emerging interest in the field.

Undersea Surveillance During the Cold War: The United States' Sound Surveillance System Versus Soviet Union's 'Strategic Bastion'

The first active device to detect submarines, called hydrophone, was developed by the Britishers in 1915. The hydrophone was put into use during World War I to detect German submarines. Detecting the German submarines had become a strategic challenge for the Royal Navy and physicists both in the UK and in Europe. In 1917, the French scientist Paul Langevin's invention of the quartz sandwich transducer led to underwater sound transmission in submarine detection (Graff, 1981). This earlier development in the field of modern ultrasonics had helped the Royal Navy to advance its ASW research in the 1920s. World War II saw a significant advancement in the development of ASW technologies. The German's diesel-powered U-boat spent a significant amount of time on the surface to recharge its batteries; therefore, ASW detection and tracking relied on the ship's surface-search radar, airborne radar, interception of radio transmission, encryption of submarine radio message, etc. The use of sonar was limited, and it was primarily used in short-range applications. But, post-World War, sonar emerged as an important means of submarine detection.

The US Navy's extraordinary efforts in underwater acoustics research and the research carried out by Massachusetts Institute of Technology (MIT) led to major technological breakthrough in developing long-range Sound Surveillance Systems to detect and track Soviet submarines and in providing vital cueing information for tactical, deep-ocean, anti-submarine warfare. The first sound surveillance system (SOSUS) stations were sited along a stretch from Barbados to Nova Scotia looking out into the North Atlantic Ocean and comprised arrays of hydrophones placed at the bottom of the ocean. The hydrophones were connected by underwater cables to shore-based processing centres. The first prototype consisted of a 333-m-long horizontal line array of 40 hydrophones laid on the seafloor at a depth of 435 m off Bahamas. After checking the working of the prototype, similar arrays were placed to cover the US east coast and then the west coast and Hawaii. The underwater geography limited SOSUS's reach in the Atlantic Ocean. The Pacific Ocean was more clear, as the SOSUS's range is dependent upon the submarine noise.

The geographical limitation in the Atlantic is due to the Mid-Atlantic Ridge; the SOSUS coverage of the eastern Atlantic required arrays on the far side of the ocean. Therefore, the United States was looking for shore-based processing units in Britain and other European countries because early SOSUS cabling could not be longer than a hundred miles. Moreover, the cables were expensive; therefore, the United States decided to establish shore stations in Britain. By the 1960s, British Isles and Eastern Atlantic had come under the coverage of SOSUSs. The collected data in Britain were later transmitted through underwater telephone cables to the United States. Maintaining those underwater cables from enemy disruption had become a major task for the United States and its allies. By the 1970s, the SOSUS covered important junctions in the Atlantic Ocean and proved to be successful in detecting Soviet SSBNs and SSNs through their sonar signatures (see Figure 3.1) (Clark, 2015). The Greenland, Iceland, England (GIUK) Gap was one of the main transit routes for the Soviet Union's submarines to reach their patrol area in the Atlantic. The NATO naval ships were led by the Supreme Allied Commander Atlantic (SACLANT), whose primary job was to maintain active ASW capability in the GIUK and surrounding areas (The UK, the High North and the North Atlantic, n.d.). The SOSUS assisted the NATO naval ships to maintain their credible deterrence against the Soviet SSBNs and SSNs (Clark, 2015).

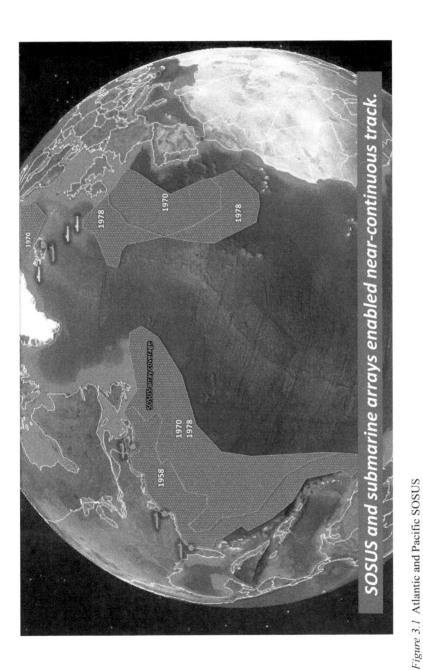

Figure 3.1 Atlantic and Pacific SOSUS

Source: (Clark, 2015) Based on data from R. F. Cross Associates, Sea-based Airborne Antisubmarine Warfare 1940–1977 (Alexandria, VA: R. F. Cross Associates, 1978).

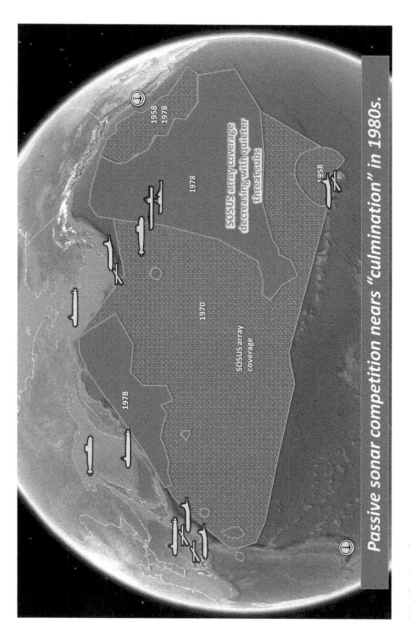

Figure 3.1 (Continued)

The success of SOSUS was the result of two major scientific developments in this field. The first development was the discovery of the deep sound channel by Maurice Ewing and J. Lamar Worzel in the 1960s. The experiment demonstrated that it is possible that low-frequency sound can travel over long distances utilizing the natural sound channels. The sound channel is a main source of the SOSUS to detect submarine noise from a long distance. The second development was the great discovery due to Bell Labs' work on Low-Frequency Analysis and Ranging (LOFAR). This technology allowed the sonar to filter out most of the sound received and focus on the low-frequency generated by submarines. The submarine-generated sounds are categorized as low frequency (narrowband), which are easily detectable by sonar from a long distance. The low-frequency sound when produced well can be distinguished from the noise at sea. The United States and its allies have collected and maintained a database of Soviet submarine's acoustic signatures, which served as a basis for the United States to distinguish the Soviet's submarines from other random noises at high seas. Later, the US Navy created the Integrated Undersea Surveillance System (IUSS) centred around the SOSUS by integrating various other sources of information related to undersea activity from surface warships, submarines, and research vessels. By 1981, the United States had deployed SOSUS in 36 installations, including facilities in the United States, the UK, Turkey, Japan, the Aleutians, Hawaii, Puerto Rico, Bermuda, Barbados, Canada, Norway, Iceland, the Azores, Italy, Denmark, Gibraltar, the Ryukyus, Panama, the Philippines, Guam, and Diego Gracia (Clark, 2015; Ball & Tanter, 2015).

The United States has spent nearly 7 US billion dollars, roughly 16 percent of the overall navy's budget in the 1980s for ASW (Wit, 1981). Substantial amount was devoted to maintain the SOSUS and underwater surveillance systems. Soviet submarines adopted a much quieter technology in their diesel submarines of the 1980s, rendering them more difficult to be detected (Ford & Rosenberg, 2005). The development of Surveillance Towed Array Sensor Systems (SURTASS) ships trailing long hydrophones capable of uplinking acoustic intelligence via satellite to the ground stations enhanced the US effort to track the submarines from a long distance. The SURTASS later upgraded to low-frequency active sonar (LFA) with long-range active sonar to detect even quieter submarines at long distances. In addition to that, the Rapidly Deployable Surveillance System (RDSS) system was developed for tactical application, which could be dropped into the ocean at a short notice and transformed into mini-SOSUS on the ocean floor. This system was used in areas where underwater surveillance was not available or not practical (The Navy's Rapidly Deployable Surveillance System Needs to Be Reassesse, 1983). During the Cold War period, the

Figure 3.2 Japan 'Fish Hook' (SOSUS)

Source: (Ball & Tanter, 2015), 'Map reproduced with the permission of CartoGIS Services, Scholarly Information Services, The Australian National University'.

SOSUSs and passive sonar provided a great advantage against the Soviet submarines.

The end of the Cold War and break-up of the Soviet Union in the 1990s resulted in major structural changes in the US ASW operations. Post-Cold War, there was a perceptible change in the nature of warfare and the priority shifted to the littorals. The US Navy submarines are now built to meet the specification to operate in the littorals congested with civilian shipping vessels and with ability to strike inland using cruise missiles. However, the US Navy's wide area of surveillance retained its position partly in the Atlantic and Pacific Ocean, due to the rise of China as a dominant naval power in the post-2000 period.

The SOSUS provided robust ASW capability to the United States and the allied forces to detect and track the Soviet SSBNs and SSNs taking deterrent patrol in the Atlantic or Pacific Ocean. But, the SOSUS is now much more evolved with the IUSS, which has integrated a wide array of surface and subsurface sensors, satellite surveillance, and state-of-the-art massive data processing storage – enabling the US Navy to store a large volume of data and allowing the operator to retrieve historical information, in addition to seeing real-time information. This allows ASW operators to check information with the previous missions or access other ships' information processed by the onshore centre. The US transition from the fixed sensors in the seafloor to sensors in mobile platforms indicates that the navy is better equipped to operate and interpret weak signals even in complex environments. The United States is also shifting to autonomous/unmanned ASW concept for long-range, long-endurance vessels, with complete autonomy; yet, it can successfully navigate and complete the mission on its own. The US Defence Advanced Research Projects Agency (DARPA) has developed two systems under the programme called 'Distributed Agile Submarine Hunting (DASH)': The Transformational Reliable Acoustic Path System (TRAPS) – a fixed passive sonar node, designed to achieve large-area coverage – and Submarine Hold at RisK (SHARK), an unmanned underwater vehicle (UUV) active sonar platform to track submarines after initial detections are made (Distributed Agile Submarine Hunting (DASH) Program Completes Milestones, 2013). The introduction of artificial intelligence/machine learning will significantly improve the detection capabilities of the manned and unmanned ASW mission in future. This would bring major structural changes to the US ASW capability by improving the accuracy and reliability in detecting, tracking, and classifying underwater objects in the future.

Soviet Union's 'Strategic Bastion'

In the early days of the Cold War, Soviet Union's emphasis was on developing SSBNs based on greater invisibility, inaccessibility, and invulnerability (McConnell, 1985). The Soviet was convinced that nuclear-powered submarines could not be readily detectable at sea, and the principal method to combat nuclear-powered submarines would be through launching missile or air strike against the naval port housing the submarine. But this perception

started to change in the 1980s, when the United States achieved technological breakthrough in ASW technology. Fearing the US naval capabilities
to detect the Soviet Russian SSBNs and SSNs, Soviet developed 'Bastion
Concept', to ensure the survival of Soviet SSBNs in near-home waters,
where they can be protected by the Soviet surface vessels and aircraft from
enemy submarines (Breemer, 2008). The term 'Strategic Bastion' refers to
adjacent seas protected by a combination of fixed sensor installations and
anti-submarine forces composed of submarines, surface ships, and aircraft
(The Russian Navy: A Historic Transition, 2015). The Soviets had created
Strategic Bastion in the Barents Sea in the north and Sea of Okhotsk in the
Pacific for SSBNs to serve as 'strategic national reserves' (The Russian
Navy: A Historic Transition, 2015). The Soviet also developed wide varieties of SSNs and diesel-electric submarines to bolster the Soviet Navy's
fight against the NATO and US forces in the region.

The 'Bastion' concept continued even in the post-Cold War environment, where Russia modernized and restructured its naval force in the
Kola Peninsula to adapt to the new security environment. In 1998, Russian
President Yeltsin declared the establishment of the North Strategic Bastion
(NSB). The main role of NSB was to provide credible nuclear deterrence
and to develop an independent naval force to secure Russian interests in
the world's oceans (Atland, 2007). During this period, the Russian Navy
revitalized seabed sensors located in the Barents Sea, the Kara Sea, the Sea
of Okhotsk, and other waters close to the Eurasian continent to monitor
the offensive patrolling carried out by the United States, UK, and French
SSBNs in the area. Furthermore, the sea-based sensors strengthened the
underwater domain awareness capability of the Russian naval force engaged
in defensive posture in the 'Bastion'. As the Arctic and Northern Sea route
gained prominence, the Russian Navy invested and developed underwater
maritime sensors similar to the US SOSUS – ready to deploy sea-based
sensors in the region.

The Russian deployment of submarines in the Atlantic and Mediterranean
during the Syrian conflict shows Moscow adapting to the new maritime
environment at a quicker rate. Using underwater surveillance systems, Russia has developed a naval system for seabed warfare. Russian Directorate
for Deep Sea Research or *GlavnoyeUpravlenieGlubokovodskIssledovanii*
(GUGI) was formed in 1976; it is a naval research organization and works
independently in building underwater defence mechanisms for Russia. The
GUGI provides Russia the required capability in the Arctic Circle and deep-
ocean region, which includes

1. Conducting subsea surveys, mapping, and sampling of the Arctic
    region and seas around Russia.

2. Placing military systems and underwater cables or retrieving items from the sea floor etc.
3. Maintaining sonar equipment and systems.
4. Developing an operational capability to deploy submarines including midget submarines, the Poseidon strategic nuclear torpedo (Lobner, 2018).

To achieve these tasks, GUGI had a range of equipment, which included *Yantar* – an ocean research vessel to identify the underwater cables. GUGI also developed the most advanced *Harpsichord* UUV, which can be carried by surface vessels as well as submarines to improve the intelligence, surveillance, and reconnaissance (ISR) capability of the Russian Navy. Among the most advanced surveillance systems developed and deployed by GUGI is the Harmony-S surveillance system.

The Harmony-S surveillance system is capable of detecting enemy ships, submarines, and even low-flying aircraft anywhere in the world's oceans. This system is termed as an autonomous seabed station (ASS) that can be launched from a submarine into the ocean floors. Once deployed, the ASS turns into a hydro-acoustic station in the seabed; the ASS has fixed multi-element horse-type sonar antennas that can perform both passive and active roles to detect and classify underwater and surface targets (Ramm, 2016). The gathered information through pop-up buoy can be transmitted through satellite to command centres. The ASS system is powered by a lithium polymer battery specially designed for underwater scenarios. Once the task is completed, the ASS antennas can be retrieved through the submarine. Russia has built special-purpose submarines to deploy and retrieve the ASS antennas from the deep oceans. Russia has also built nuclear-powered *Belgorod* and *Khabarovsk* submarines designed to carry the ASS systems. The sensors are deployed in Russia's Northern Sea Fleet at Olenya Guba near Murmansk and in the Baltic Sea. Russia has also deployed these capabilities in the Arctic region, foreseeing a western naval interest in the region.

The United States and Soviet adopted different strategies based on their geographical location and materialistic capability to build SSBN and SSN programme to achieve credible nuclear deterrence. Detecting and tracking enemy submarines remained an elusive concept during the Cold War period. But, the US Navy's perseverance in building an effective ASW strategy against the Soviet led to the development of the SOSUS. The SOSUS provides assistance to the US and NATO's navies to keep track of the Soviet's SSBNs and SSNs venturing out to the open oceans. The United States' 'Offensive ASW' strategy had significantly affected the Soviet's naval planner in the 1970s. This forced the Soviet Union to adopt the 'bastion' concept to escape the United States' containment strategy in the Atlantic and

Pacific oceans. Moreover, the Soviet's submarine-launched ballistic missiles (SLBMs) can reach any location in the United States from the Arctic Sea, close to the Russian shore, making the 'bastion' strategy less expensive and at the same time providing a safer place for the SSBNs to operate. The Soviet maintained the highest number of SSNs and diesel-electric submarines (SSKs) to disrupt the US and NATO Navy's activity in the GIUK Gap and track the NATO navies in the Russian coastal waters. Russia's 'Defensive ASW' strategy does ensure the SSBNs to use the ocean to conduct necessary military operations to defeat or deter enemy's submarines/nuclear threat.

Even in the post-Cold War era, the United States and Russia remain actively engaged in anti-submarine warfare. Even though Russia was facing a severe financial crisis in the aftermath of Soviet Union disintegration in the 1990s, it retained some of the SSBNs and SSNs, if not all, to maintain credible deterrence against the United States. It has rebuilt its naval fleet with the introduction of the new Typhoon-class submarines and newer submarines to maintain its parity with the US nuclear and conventional capabilities. The United States' investment in undersea surveillance doubled during this period and has developed more advanced, sophisticated, manned and unmanned surveillance platforms to monitor the undersea areas.

Underwater Surveillance Systems of China and Japan

The United States' advancement in the ASW technological domain has put various other countries in a state of perplexity. The Asian maritime powers like China and Japan have also built underwater surveillance systems to gather information of enemy submarine activities in the home waters. Japan is a close ally of the United States in the Indo-Pacific region and has developed underwater surveillance systems around the archipelago, independent of the United States, to monitor Chinese and North Korean submarines in their water. This section illustrates the undersea surveillance capability of the Chinese and the Japanese, which have a different perception on the undersea environment.

China

Underwater surveillance is emerging as one of core aspects of Chinese naval power. Chinese ASW strategy adopts a mixture of both the 'bastion strategy' and 'offensive ASW Strategy'. The Chinese control over the South China Sea through military infrastructure in Spratly and Parcel Island provides the PLA Navy strategic advantage to convert the entire South China Sea into a 'bastion'. According to Asia Maritime Transparency Initiative

(AMTI) reports, China has 20 military outposts in the Paracel Islands and 7 in the Spratly Islands (CHINA ISLAND TRACKER, n.d.). The occupation of these island chains in the South China Sea is a strategic move to secure SSBNs' passage to the Pacific Ocean. China has also built air stations in both islands, to support the troops, and to control high-frequency radars, jamming equipment, surface-to-air missiles HQ-9, and anti-ship missiles like YJ-12 and YJ-62 in the islands chains (United States-China Economic And Security Review Commission, 2020). The top US Commander in the Indo-Pacific, Admiral Philip Davidson, says China turned the South China Sea into the 'great wall of SAM [Surface to Air missiles]' (McLeary, 2018). These efforts would significantly boost the PLA Navy's second-strike capability.

China's offensive ASW strategy primarily involves the PLA Navy's modern class destroyers, frigates, aircraft, and helicopters. Since early 2010, most of the advanced destroyers and frigates are fitted with hull-mounted sonar, lightweight torpedo launchers, and rocket-launched depth chargers (Joe, 2018). However, such systems indicate that it is useful only in short-range detection and engagement purposes only. The newer-class destroyers (Type-055 and Type 052D), frigates (Type-054A), and corvette (Type-056/A) are equipped with towed array sonar system (TASS) and variable-depth sonar (VDS) systems (Joe, 2018). China is also planning to replace the old assets with the Type-055 and Type-052 destroyers; this would significantly improve PLA Navy's focus on the anti-submarine warfare. The Frigate Type-054A, the workhorse of PLA Navy is also equipped with passive sonars and variable-depth sonars. China has also managed to build huge number of corvettes to replace the small patrol boats for coastal defence. The previous version of the Type-056 corvette was not equipped

Table 3.1 Principle Surface Vessels in ASW Role

Vessels/aircraft/helicopter	In service	ASW systems
Renhai (Type 055) class (DDGHM)	3	Bow-mounted, towed sonar suite (passive line and active VDS)
Luyang III (Type 052D) class (DDGHM)	17	Hull-mounted; towed sonar suite (passive line and active VDS)
Jiangkai II (Type 054A) class (FFGHM)	30	Hull-mounted; Type 206 passive towed array or Type 311 active/passive variable-depth sonar (from Huanggang)
Jiangdao (Type 056A)	50	Bow-mounted sonar. VDS and SJG-206 towed array (ASW variant)

Source: Based on the IISS Military Balance (Submarines and Sub-Surface Warfare, 2021), Janes Fighting Ship (Destroyers, 2021), (Frigates, 2021), (Corvette, 2021).

Table 3.2 Principal ASW Helicopter and Maritime Patrol Aircraft

Helicopter/aircraft	In service	ASW systems
HAIG Z-9C	22	Thomson-Sintra HS-12 or Type 605 dipping sonar
KA-28PL	14	Chin-mounted surface-search radar, sonobuoys, MAD, and dipping sonar
Z-18F	6	Dipping sonar, can carry up to 32 sonobuoys
Y-8Q (Maritime Patrol Aircraft)	Approx. 17	Surface-search radar, MAD, and four sonobuoy launchers

Source: Based on the IISS Military Balance (Submarines and Sub-Surface Warfare, 2021), Jane's Fighting Ship (Shipborne Aircraft, 2021).

with ASW suites but the upgraded model, the Type 056A, presents a towed array sonar and variable-depth sonar. The Chinese PLA Navy also field in helicopters HAIG Z-9C, KA-28PL, and Z-18F for the ASW role. The PLA Navy has also inducted Y-8Q first independently developed and manufactured fixed-wing anti-submarine patrol aircraft (China's first anti-submarine patrol aircraft in 60 seconds, 2021). The Y-8Q resembles the US P-3 Orion and is equipped with modern sensors and weapon suits similar to other maritime patrol aircraft. There is a firm requirement from China for long-range maritime patrol aircraft. The Y-8Q exactly fits the description of PLA Navy's requirements for routine patrol of crucial sea lanes in the East and South China Sea. It is highly likely that all the three naval theatre commands may operate Y-8Q MPA.

China has largest maritime force on the globe with about 355 vessels. It has planned to increase the fleet strength to 460 by the end of 2030 to match the US Navy (Shelbourne, 2021). China is not only focusing on numerical strength but also investing on superior built quality to compete with the US naval forces in the Indo-Pacific region. While China has already achieved significant capabilities for coastal defence, the long-term objective of the PLA Navy is to achieve blue-water capabilities and expand underwater surveillance capability in the Western Pacific and South China Sea.

But the major discussion on the Chinese plans for underwater surveillance systems started in 2016 when the South China Morning Post reported that one of the China State Shipbuilding Corporation proposed to the central government about building a network of submarine detectors in the South China Sea (Wong, 2016). This news was broadcast by several western media about a possible Chinese plan of building underwater surveillance systems, popularly known as 'Underwater Great Wall'. However, there seems to be no strong evidence to suggest that China is building a SOSUS-like system in the South China Sea. But, China – aware of the growing importance of surveillance in the ocean for its long-term plan – launched a project called 'Sea Floor Long-Term Observatory Network Experimental Key technology of Nodes', under the 11th Five-Year plan (Xin, 2013). In 2009, Tongji University helped to set up East China Sea Seafloor Observatory in Xiaoqushan. In 2013, the Sanya submarine observation network system laid a strong foundation for the seabed observation network construction in China. In 2016, China installed the South China Sea Seafloor Observation Network (SCSSON), primarily to study dynamic and biogeochemical processes, to prevent disasters, etc. (Lu, Zhou, Peng, Yue, & Wang, 2015). The complete network was built with optical fibre, in order to meet the long-term, real-time observation requirements of China (Liu et al., 2019). This shows that China's marine scientific research is aiming to build real-time seabed observatory systems in the regional waters.

As part of the 'Blue Ocean Information Network', China Electronics Technology Group Corporation (CETC) developed an all-weather information network node system for the protection of island reefs, maritime research, and other activities. The system has two variants: a floating Integrated Information Platform (IIFP) and a more powerful Island Reef-Based Integrated Information System (IRBIS) (Zhuo, 2019). In 2019, Asia Maritime Transparency Initiative reported that China deployed five IIFPs around Hainan and one IRBIS fixed platform at Bombay Reef (Exploring China's Unmanned Ocean Network, 2020). These systems are most likely to use as an early warning system to send warnings in case of seaquakes, earthquakes, and tsunami in the region. At the same time, it has significant military uses like tracking of ships and submarines in real time and continuous surveillance of offshore targets. China's State Oceanic Administration is undertaking scientific and technical research. It is also actively participating in gathering underwater chemical, biological and other property information of the East and South China Sea, which is crucial for China to operate its submarine safely and effectively in the region.

China is also expanding its underwater surveillance beyond its first island chain. China established a listening post near the US naval base in Guam. According to the South China Morning Post report, the listening post was planted in the Challenger Deep of the Mariana Trench and another one near Yap, an island in the Federation States of Micronesia. The underwater sensor located between Guam and Palau would provide China an advantage to keep an eye on the US submarine force in the western Pacific. China is also foraying into the Indian Ocean for collecting oceanic data related to chemical, biological, and underwater temperature for military purposes. As China is closing its gap with the Indian Ocean, it might even set up a listening post in the Indian sub-continent either in Sri Lanka or in Pakistan to track Indian submarines and surface ships in the region.

Japan

During the Cold War, Japan had created underwater surveillance systems for monitoring Soviet and North Korean submarines and tracking enemy's surveillance vessels in the Pacific Ocean. Japan Maritime Self-Defence Force (JMSDF) has put in place a comprehensive architecture for underwater surveillance, which was integrated with JMSDF Ocean Surveillance Information System Evolutionary Development (JOED) system at the JMSDF fleet headquarters at Yokosuka (Ball & Tanter, 2015). Japan now has eight signal intelligence (SIGINT) Circularly Disposed Antenna Array (CDAA) for both signal interception as well as HF and DF activity. The surveillance systems were deployed across Japan in the defence harbour, internal and adjacent straits (see Figure 3.3). Japan also augmented its defence capability by building strong defence cooperation with the United States. The US military presence in Japan and regular interaction between the two armed forces enhanced Japan's naval capabilities over a period of years by monitoring and tracking Soviet submarines in the Sea of Japan.

The JMSDF operates several ocean surveillance programmes and operates two SURTASS (Surveillance Towed Array Sensor System) acoustic measurement (or underwater listening) ships. In 2020, Japan launched its ocean surveillance vessel, '*Aki*' – a *Hibiki*-class ocean surveillance vessel, but the first since its sister ship was launched in 1992 (Ryall, 2020). Japan operates the largest fleet of maritime surveillance aircraft in the world, after the US Navy, comprising P-3C Orion long-range maritime patrol aircraft and SH-60J Seahawk patrol helicopters. Japan's Coast Guard vessels were also involved in surveying the Japanese coast and other hydrographic surveys of the sea lane adjacent to the Japanese coasts. Starting from the 1950s to now, JMSDF and Japan Coast Guard have created a comprehensive underwater surveillance network along the Japanese coast against the Russian, Chinese, and North Korean submarine threats.

The brief overview of the undersea surveillance network capabilities of select countries in the Indo-Pacific shows that China and Russia have built comprehensive undersea surveillance systems. Japan, an important ally of the United States in Asia, also built a SOSUS-equivalent undersea surveillance unit to protect its underwater interests from the North Korean midget submarines, and the Chinese and Russian submarine force. The Russian and Chinese deployment of the underwater autonomous vehicle for ASW purposes will hype the underwater activities in the Western Pacific and East and South China Sea. China's underwater drones and underwater surveillance

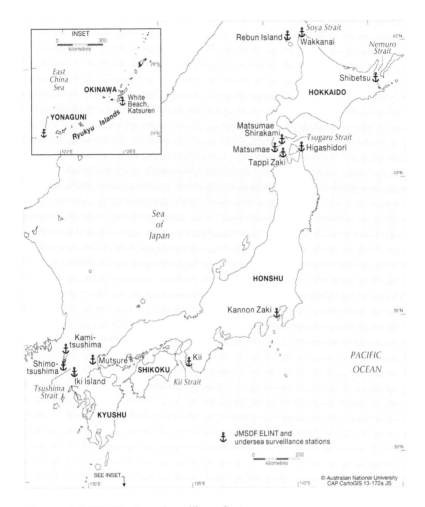

Figure 3.3 JMSDF Undersea Surveillance System

Source: (Ball & Tanter, 2015), 'Map reproduced with the permission of CartoGIS Services, Scholarly Information Services, The Australian National University'.

network in the South China Sea would enable it to operate the SSBNs and SSNs safely. The Indian Navy – which has a huge stake in the regional maritime security interest in the region – is trying to expand its operational envelope in the region and is concerned with the Chinese surface and sub-surface sensors and sonars operating in the region.

In conclusion, the US SOSUSs are primarily aimed at creating wide-area underwater surveillance to detect, track, and classify the Soviet SSBNs and SSNs. The threat of Soviet SSBNs and SSNs was significantly reduced in the post-Cold War era. Russia now operates only a few SSBNs and SSNs compared to the Cold War era. The post-Cold War maritime environment has also changed and naval warfare has moved close to the global littorals. The diesel submarines and AIP technology now pose a major challenge to underwater surveillance. Global politics has also shifted from the Atlantic to the Indo-Pacific as China has emerged as a major maritime power in the twenty-first century.

Given the cost and practical difficulties to maintain an underwater cable network, it might be an expensive task for a country like India. Even, round-the-clock monitoring of important chokepoints in the Indian Ocean – Malacca, Lombok, and Sunda straits – is a mammoth task for the Indian Navy as well other agencies involved in the deployment and maintenance of the array of sonar in the seafloor. The area is congested with maritime traffic as a huge flow of shipping traffic passes through crucial chokepoints in the IOR, to various destinations in the Pacific; moreover, the presence of huge fishing vessels makes the deployment or maintenance of array of sensors close to chokepoints not only a difficult task but also highly vulnerable to disruption from either enemy action or fishing activity. Since some of these crucial chokepoints are under the control of the Indonesian government, any clandestine attempt to deploy an array of sensors in the sea floors of Indonesia's Exclusive Economic Zone (EEZ) would be viewed as an aggressive act; furthermore, it would jeopardise India's relationship with Jakarta. Since Indonesia itself operates submarines in this area, Jakarta will not allow its littorals to be dominated by the Indian Navy.

Bibliography

Atland, K. (2007). The Introduction, Adoption and Implementation of Russia's "Northern Strategic Bastion" Concept, 1992–1999. *Journal of Slavic Military Studies*, 499–528.

Ball, D., & Tanter, R. (2015). *The Tools of Owatatsumi Japan's Ocean Surveillance and Coastal Defence Capabilities*. Canberra: ANU Press.

Breemer, J. S. (2008). The Soviet Navy's SSBN Bastions: Why Explanations. *The RUSI Journal*, 33–39.

China Economic and Security Review Commission (2020). *U.S.-China Economic and Security Review Commission*. Washington: U.S.-China Economic and Security Review Commission.

China Island Tracker. (n.d.). *Asia Maritime Transparency Initiative*. Retrieved from: https://amti.csis.org/island-tracker/china/#Spratly%20Islands (Accessed on June 2, 2021).

China's First Anti-Submarine Patrol Aircraft in 60 Seconds. (2021, November 24). China's First Anti-Submarine Patrol Aircraft in 60 Seconds. *CGTN*. Retrieved from: https://news.cgtn.com/news/2021-11-24/China-s-first-anti-submarine-patrol-aircraft-in-60-seconds-15rCVOKKhhK/index.html (Accessed on November 25, 2021).

Clark, B. (2015). *The Emerging Era in Undersea Warfare*. Washington: The Center for Strategic and Budgetary Assessments.

Corvette. (2021). *Janes Fighting Ships*. Retrieved from: https://en.wikipedia.org/wiki/Corvette.

Department of the Navy. (1992). *Forward From the Sea*. Washington: Department of the Navy.

Destroyers. (2021). *Janes Fighting Ships*. Retrieved from: https://en.wikipedia.org/wiki/Jane%27s_Fighting_Ships.

Distributed Agile Submarine Hunting (DASH) Program Completes Milestones. (2013, March 4). *DARPA*. Retrieved from: www.darpa.mil/news-events/2013-04-03 (Accessed on May 11, 2021).

Exploring China's Unmanned Ocean Network. (2020, June 16). *Asia Maritime Transparency Initiative*. Retrieved from: https://amti.csis.org/exploring-chinas-unmanned-ocean-network/ (Accessed on June 2, 2021).

Ford, C. A., & Rosenberg, D. A. (2005). The Naval Intelligence Underpinnings of Reagan's Maritime Strategy. *Journal of Strategic Studies*, 379–409.

Frigates. (2021). *Janes Fighting Ship*. Retrieved from: https://www.janes.com/defence-news/naval-weapons.

Graff, K. F. (1981). A History of Ultrasonics. In W. P. Mason & R. Thurston (eds.), *Physical Acoustics Principles and Methods* (pp. 1–97). London: Academic Press.

Joe, R. (2018, September 12). The Chinese Navy's Growing Anti-Submarine Warfare Capabilities. *The Diplomat*. Retrieved from: https://thediplomat.com/2018/09/the-chinese-surface-fleets-growing-anti-submarine-warfare-capabilities/ (Accessed on November 11, 2021).

Liu, L., Liao, Z., Chen, C., Chen, J., Niu, J., Jia, Y., . . . Liu, T. (2019). A Seabed Real-Time Sensing System for In-Situ Long-Term Multi-Parameter Observation Applications. *Sensors*, 19.

Lobner, P. (2018, May 21). *You Need to Know About Russia's Main Directorate of Deep-Sea Research (GUGI)*. The Lyncean Group of San Diego. Retrieved from: https://lynceans.org/all-posts/you-need-to-know-about-russias-main-directorate-of-deep-sea-research-gugi/ (Accessed on May 1, 2021).

Lu, F., Zhou, H., Peng, X., Yue, J., & Wang, P. (2015). Technical Preparation and Prototype Development for Long-Term Cabled Seafloor Observatories in Chinese Marginal Seas. In P. Favali, L. Beranzoli, & A. D. Santis (eds.), *Seafloor Observatories* (pp. 503–529). Berlin, Heidelberg: Springer.

McConnell, J. M. (1985). New Soviet Methods for Anti-Submarine Warfare. *Naval War College Review*, 16–27.

McLeary, P. (2018, November 17). China Has Built 'Great Wall of SAMs' In Pacific: US Adm. Davidson. *Breaking Defense*. Retrieved from: https://breakingdefense.com/2018/11/china-has-built-great-wall-of-sams-in-pacific-us-adm-davidson/ (Accessed on June 15, 2021).

Naval Doctrine Command. (1998). *Littoral Anti-Submarine Warfare Concept.* Naval Doctrine Command. Retrieved from: https://man.fas.org/dod-101/sys/ship/docs/aswcncpt.htm.

The Navy's Rapidly Deployable Surveillance System Needs to Be Reassesse. (1983). *The Navy's Rapidly Deployable Surveillance System Needs to Be Reassesse.* Report by the Comptrollergeneral of the United States, Washington.

Office of Naval Intelligence. (2015). *The Russian Navy: A Historic Transition.* Washington: Office of Naval Intelligence.

Pittman, J. C. (2008, Feb). *Zone Defense – Anti-Submarine Warfare Strategy in the Age of Littoral Warfare.* U.S. Army Command and General Staff College. Retrieved from: www.hsdl.org/?view&did=729684 (Accessed on April 11, 2021).

Ramm, A. (2016, November 30). Russian 'Harmony' for Maritime Surveillance. *Russia Beyond.* Retrieved from: www.rbth.com/economics/defence/2016/11/30/russian-harmony-for-maritime-surveillance_652217 (Accessed on May 15, 2021).

Ryall, J. (2020, March 11). Japan Builds New Surveillance Warship Targeting Chinese, North Korean Submarines. *South China Morning Post.* Retrieved from: www.scmp.com/week-asia/politics/article/3074680/japan-builds-new-surveillance-warship-targeting-chinese-north (Accessed on August 2, 2021).

Shelbourne, M. (2021, November 3). China Has World's Largest Navy With 355 Ships and Counting, Says Pentagon. *USNI News.* Retrieved from: https://news.usni.org/2021/11/03/china-has-worlds-largest-navy-with-355-ships-and-counting-says-pentagon (Accessed on November 11, 2021).

Shipborne aircraft. (2021). *Janes Fighting Ships.* Retrieved from: https://shop.janes.com/fighting-ships-yearbook-20-21-6541-3000008248.

Submarines and Sub-Surface Warfare. (2021). *Military Balance IISS*, ci. Retrieved from: https://www.tandfonline.com/doi/abs/10.1080/04597222.2020.1869450.

The UK, the High North and the North Atlantic. (n.d.). *UK Parliament.* Retrieved from: https://publications.parliament.uk/pa/cm201719/cmselect/cmdfence/388/38806.htm (Accessed on February 27, 2021).

Wit, J. S. (1981, February). Advances in Antisubmarine Warfare. *Scientific American*, 31–41.

Wong, C. (2016, May 19). Underwater Great Wall': Chinese Firm Proposes Building Network of Submarine Detectors to Boost Nation's Defence. *South China Morning Post.* Retrieved from: www.scmp.com/news/china/diplomacy-defence/article/1947212/underwater-great- (Accessed on October 12, 2021).

Xin, L. (2013). The East China Sea Floor Observatory Network Systems Engineering: Economic Analysis and Prospects. In *Proceedings of the 2013 International Conference on Advances in Social Science, Humanities, and Management* (pp. 711–718). Amsterdam: Atlantis Press.

Zhuo, C. (2019, April 1). China Launches New System to Defend Islands and Reefs in South China Sea. *PLA Daily.* Retrieved from: http://english.pladaily.com.cn/view/2019-04/01/content_9464939.htm (Accessed on October 11, 2021).

4 Emerging Era in Anti-Submarine Warfare

In the post-World War era, the evolution of nuclear-powered submarines (SSNs) and Ballistic Missile Submarines (SSBNs) to complete the nuclear triad gained importance, primarily in Russia and the United States, followed by Britain and France. The SSBNs completed the important underwater leg of the triad, simultaneously fulfilling the role of deterrence and second retaliatory strike. Advancements in technology became important requirements to gather intelligence and to infer the adversary's intents and plans. These in turn translated to trailing adversary maritime fleets without being detected, eavesdropping on their movement, and gathering data on speed, operational parameters, as well as to characterise their acoustic signature. Technology gains led to quieter submarines and nuclear-powered long-endurance submarines with a variety of armaments.

These features, while enhancing the offensive role of submarines, also added to the difficulty in planning and carrying out anti-submarine warfare operations. Challenges to primary Anti-Submarine Warfare (ASW) functions to detect (stealthier and quieter), track (without being detected), and communicate (without getting compromised) increased manifold. Detecting, tracking, and gathering vital intelligence of submarines remain a key part of any country's undersea warfare programme. To this end, the United States and Russia have adopted distinctively different strategies.

The United States developed what is known as 'Strategic ASW', a strategy applied as an offensive measure to delay the enemy submarine's intrusion and reach to the operation or patrol area. The aim here was to suppress the enemy submarine's activity and ASW measures would be enforced. The deployment of transatlantic Sound Surveillance Systems (SOSUSs), Electronic Intelligence (ELINT), electro-optical/infrared satellites, and towed array sensors for covering the Greenland, Iceland, England (GIUK) gap is a classic example of 'Strategic ASW'. The combination of underwater sonars, sensors, and aerial/land-based radar systems is engineered to provide

DOI: 10.4324/9781003298380-4

round-the-clock surveillance of adversary submarine movement and thus forms an important deterrent element. Russia adopted a somewhat localized version of the ASW technique, called 'defensive ASW' to deter enemy's naval vessels from entering the operational areas of Russian submarines. The *Bastion Concept*, adopted by Russia, identified a well-defended area for nuclear submarine operations. The defined space could be large and patrolled/defended by fighting vessels of the Russian naval fleet and shore-based aircraft. The Russian forces deployed SSNs for long-range patrol and diesel-electronic submarines for monitoring the undersea movement in the Barents sea. This strategy not only helped the two nations to maintain the credible nuclear deterrence but also enhanced and advanced underwater sonar, sensors, and area-denial weapons.

In both scenarios, technology played a decisive role. The end of the Cold War impacted the strategic approach – a cash-starved Russia shelved many of its undersea warfare research projects and the United States too downsized its ASW research. The United States has a string of laboratories devoted to marine research, of which underwater and seabed study forms an important component. Research institutes like the Woods Hole Oceanographic Institution, Scripps Institute of Oceanography, University of California, San Francisco State University, the Monterey Bay Aquarium Research Institute, Space and Naval Warfare (SPAWAR) Systems Center Pacific, and the Office of Naval Research were leading scientific inquiry in marine research and also played a huge part in the nation's defence and developing new technologies for the US Navy. Moreover, the United States' transatlantic joint initiative with UK, France, and Germany, such as the SACLANT ASW Research Center (called SACLANT Undersea Research centre from 1987) started in 1959 and continued its operations even in the post-Cold War era, until 2003. It was renamed as the Centre for Maritime Research and Experimentation (CMRE) to explore undersea marine environments and is still helping NATO powers to understand the underwater environment in the Atlantic Ocean/Mediterranean Sea. The sharing of knowledge between allied navies helped the United States and NATO to successfully design and develop underwater surveillance systems to detect, track, and classify the submarines. The scholarship in the field of undersea research has also transformed the European defence enterprise to innovate and make sonar for ASW operations in the European Waters. In the United States, the focus is shifting from a platform-centric approach to a domain-centric approach. The emergence of off-board undersea systems like UUVs, UAVs, and Distributed Netted Systems (DNS) has completely changed the navy's ability to understand what's happening underwater; the US Navy wants to connect the information from the off-board sensors with the onboard sensors to create common operational picture (COP) (Design for Undersea Warfare Update One, 2012). As the warfare is moving close to the littorals of the Indo-Pacific region, the undersea warfare remains a top priority for the US Navy. This chapter will focus on the latest discussions in the future of undersea warfare and how UDA is gaining importance in understanding the subsurface challenges in the Indian Ocean.

The New Era of Undersea Warfare

In the twenty-first century, undersea warfare is changing and the role of submarines and ASW is also adapting to the new maritime environment. Dr. Norman Friedman, a well-known naval analyst, published a report called *Strategic Submarines and Strategic Stability: Looking Towards the 2030s* in 2020; it provides a clear overview of the future course of strategic submarines and ASW operations. His primary focus is on the development of strategic submarines and its importance in global warfare. Friedman claims that understanding the ocean would provide an opportunity for SSBNs and SSNs to find a perfect place to hide deep inside the oceans, making the detection capability weak. Now, the very large SSBNs and SSNs are capable of carrying decoy, torpedo defence systems on a much larger scale than the small ones. Moreover, the increasing commercial activity at sea would make detection difficult in both the high seas and the littorals. Dr. Friedman categorizes the ASW into tactical and strategic forms and divides it into three different approaches to ASW. The first approach focuses on the enemy submarines (it is called offensive ASW). This approach uses long-range sensors to track the hostile submarine and the concentration of ASW forces in important chokepoints. The first approach aims to deter the hostile submarine from attacking, making concentrations of targets (convoys or battle groups) that would discourage the enemy from launching the attack. The first approach aims to protect the target, rather than destroy the submarine. This includes evasion and defensive measures such as anti-torpedo weapons (Friedman, 2019). Even if the submarine was detected successfully, the ultimate question is, whether it can be destroyed? Because it requires more precise positioning of the enemy submarine and overcoming the defensive measures in this modern day is a difficult task. Dr. Friedman says that policymakers would rather choose to invest in far more important naval missions like power projects than ASW because it is difficult to judge its success during peacetime. He also suggests that it is far more profitable to target and attack strategic submarine systems such as the Very-Low Frequency (VLF) station's support facilities rather than the submarine itself.

Another important work on ASW comes from Captain (retd.) William J. Toti, Commander of the USS Indianapolis (SSN-697), Submarine Squadron

3 in Pearl Harbor, and Fleet Antisubmarine Warfare Command detachment Norfolk. Captain (retd.) Toti authored a new doctrine titled *Full Spectrum of ASW;* it argues that many of the ASW operation successes were transitory, and in most of the cases navies have no clue how they defeated hostile submarines. Captain Toti observed this during 2004 RIMPAC exercise, deconstructing the perception of modern ASW as follows:

> the ASW activity was focused on detecting, localizing, and 'destroying' enemy submarines, many of the detection tactics that consumed nearly all of the friendly forces' ASW efforts simply did not work. Most of the submarines remained undetected. Yet friendly forces were still achieving transitory ASW successes against submarines they had not even detected. In other words, victory could be achieved through serendipity.
>
> (Toti, 2014)

So, Captain Toti developed a strategy by converting the 'serendipity' into 'effectivity' and translated these accidental methods into deliberate tactics. He claims that 'the 'sensor uber alles' approach, although still regaled by the Cold Warriors, was no longer affordable nor effective' (Toti, 2014). The 'Full Spectrum ASW' focuses on the fundamental aspects of submarine operations. In his research, he points out that both submarine operations and ASW operations have increasingly become complex and a difficult task; at the same time, like all human beings the submariners also react to outside stimuli. Therefore, he proposed in his doctrine that it is possible to shape a submarine's behaviour even when it hasn't been detected. This has led to the precepts of ten commandments to achieve Full-Spectrum ASW (Medcalf, Mansted, Frühling, & Goldrick, 2020).

Table 4.1 Full-Spectrum ASW

---

1. Create conditions where an adversary chooses not to employ submarines
2. Defeat submarine in port
3. Defeat submarines' shore-based command and control (C2) capability
4. Defeat submarines near port, in denied areas
5. Defeat submarines in chokepoints
6. Defeat submarines in open ocean
7. Draw enemy submarines into ASW 'kill boxes', to a time and place of our choosing
8. Mask our forces from submarine detection or classification
9. Defeat the submarines in close battle
10. Defeat the incoming torpedo

Source: Captain Toti's Full Spectrum ASW.

The ten commandments suggest that suppressing the enemy submarine from reaching its operation area is a prime focus of the 'Full-Spectrum ASW'. The engagement with the enemy submarine was discussed as a last resort. Capt. Toti suggests the only effective form of ASW is to stop the proliferation of submarines. He says that the submarines may provide some tactical advantage, but it would cause much more severe damage to the nations that employ them. Capt. Toti's idea of 'Full-Spectrum ASW' is based on a combination of strategic and psychological warfare tactics to discourage the adversaries from using submarines. All the major scholars in the field indicate that submarine warfare is undergoing profound changes in the post-Cold War period. Particularly, technology plays a dominant role in submarine operations.

The technological advancements in acoustic domains are bringing about revolutionary changes in the ASW philosophy. The progress in non-acoustic technologies may add advantage to the ASW strategy in the future. The changes are marked across a host of functions and in particular relate to wireless transmission, unmanned underwater vehicles, supercomputers, AI, and big data. Among these, UUVs can be considered to be more impactful and are game changers in a way, as they can be adapted to a number of diverse missions. Some examples include functions like

1. Intelligence, Surveillance, and Reconnaissance (ISR)
2. Mine Countermeasures (MCM)
3. Anti-Submarine Warfare (ASW)
4. Inspection/Identification
5. Oceanography
6. Port Security
7. Communications/Navigation Network Node (CN3)
8. Payload Delivery
9. Information Operations
10. Time Critical Strike (TCS) (Button, Kamp, Curtin, & Dryden, 2009).

The autonomous systems are economical as they do not have to cater to crew requirements and hence are less sophisticated and cheaper to build (Clark, Cropsey, & Walton, 2020). In addition to being faster to build, they can employ the latest technologies. These new autonomous systems, deployable sonars, automated acoustic processing and communications networking, could provide the navy capabilities to suppress the enemy submarine with greater effectiveness at lower costs. This would also reduce the burden on manned platforms like DDGs, P8-A, and SSNs, and allow greater flexibility to use the manned platform for other high-priority operations. With such attractive attributes, UUVs will be in demand by other navies also, and an undesirable fallout will result in unregulated traffic in an already congested and contested underwater domain.

By 2030, the US Navy will have a significant number of autonomous vehicles in its fleet and could be deployed for 40 distinct missions. The focus will be on the high-priority missions, which reinforce 'America's undersea edge' (Clark, 2015). Given its advantage in undersea warfare, research in seabed and oceanography, naval culture, and operational competence compared to their peers gives the United States the 'first mover' advantage.

Major NATO countries are following their own development agenda for unmanned vehicles and systems. The UK has unveiled a futuristic concept design called 'Atlantis' – a 'Hybrid Underwater Capability', to be built by 2040. The design calls for adaptation of a semi-autonomous mothership submarine with capability to deploy and retrieve underwater vessels (UUVs) from inaccessible areas. This project is similar to the US DARPA's Mata Ray project initiated in 2020, which has the capability to operate on long-duration, long-range missions in ocean environments (DARPA Selects Performers to Advance Unmanned Underwater Vehicle Project, 2021). In most of the cases, strategic deterrence and lethality remain the priority in undersea warfare (Eckstein, 2019). The United States' and NATO countries' orientation in undersea warfare is dominated by nuclear submarine threats.

Effective employment of sophisticated weapon systems requires unprecedented coordination and shared understanding of the underwater domain awareness. This necessitates the building of situational awareness in the undersea, which requires a constant flow of timely information from many sources ranging from oceanographic sensors to environmental models to human intelligence, and subject matter experts (Aldinger & Kao, 2005). Lt. Cdr. Finch of the Canadian Navy further conceptualizes the term UDA as an aggregation of maritime undersea monitoring strategies, processes, and data relating to the following (Finch, 2011):

1. Geophysical activity of the earth's crust for tsunami warning.
2. Maritime industrial exploration and exploitation efforts.
3. Security/defence monitoring and assessment to track the threat posed by submarines, mines, and the employment of undersea systems by transnational agents – for example criminal gangs – seeking to avoid detection.

There are limitations in the MDA as underwater threats are growing and require a more comprehensive framework to understand what is occurring in the ocean from the surface through the seabed at all times. UDA can be a useful system to monitor not only the subsurface threats but also marine life and undersea environments, and to monitor and assess the geophysics, anthological activities, and biological health of the ocean. Cdr. Arnab Das, Indian naval officer and researcher working on the UDA, has also proposed a similar approach to the UDA framework, where he suggests an integration between different aspects such as the corporate sector, marine environment, marine science and research, and national security. The core of the UDA framework is based on acoustic capacity and capability building. But, the latest innovation in non-acoustic sensors like Magnetic Anomaly Detection (MAD), using hydrodynamic techniques and Light Detection and Ranging (LIDR) Technology may also find use in the UDA framework to detect the submarine's disturbance on the ocean surface. The future growth of this technology would enable major powers to deploy these sensors on space, aerial, and surface ships for round-the-clock monitoring of the undersea maritime environment. The UDA is critically important for India, which is seeking to build underwater surveillance capability to monitor subsurface threat in the region.

Maritime/Underwater Domain Awareness: Assessing the Role for India

The MDA concept particularly gained attention in the post-9/11 attack on the United States, where the terrorists hijacked the American airplanes and crashed into the World Trade Center and Pentagon in the United States and disrupted global trade for several months. The attack on the United States has highlighted the need for a more comprehensive strategy to detect asymmetric threats. The US military and homeland security experts also believe that the lack of situational awareness led the jihadists to hijack the plane on US soil. This has led the US Homeland Security to develop a comprehensive 'National Strategy for Maritime Security' in 2005 to enhance effective cooperation between different state departments (National Plan to Achieve Maritime Domain Awareness, 2005). Rear Admiral Joseph L Nimmich, US Coast Guard and Captain Dana A. Goward, US Coast Guard (Ret.) explains the reason the MDA became important in the post-9/11 attacks. It was because the US aviation systems and pre-existing architectures, provided real-time visibility of total US aerospace and effective communication throughout the aviation systems, which allowed for 5,000 aircraft to safely land within 2 hours after the attacks. New policies were introduced to mitigate 9/11-style attacks in the future (Goward, 2008). However, the maritime domain doesn't yet have a system equivalent to civil aviation. The AIS fitted vessels have become mandatory for the larger ships; however, the shipping sector does not have robust tracking of vessels/subsurface vessels like the airline industries have developed. Again, the threat in the maritime domain is not from the ship or vessels, it may be from the cargo/containers;

so, tracking the containers or hazardous cargo is a major challenge for port authorities. That's why the United States introduced Container Security Initiative (CSI) to scan all containers bound to the United States. This combined effort of the United States in the early twenty-first century has led to a major discussion on the maritime domain awareness as a whole process to mitigate the challenges in the maritime domain.

The National Strategy has eight plans, but achieving maritime domain awareness is considered as one of the prime objectives to effectively use maritime information in support of maritime safety and security. The National Strategy for Maritime Security defined MDA as "the effective understanding of anything associated with the global maritime domain that could impact the security, safety, economy, or environment of the United States" (Goward, 2008). Even though the focus on MDA started in the United States, many western navies have similar concerns regarding the growing challenges in the maritime domain. The information-sharing/dissemination of information/data has become a key feature of the MDA. What makes the MDA successful is not collection of information or processing of information but making it available to relevant stakeholders.

The United States' MDA framework is useful to understand the global security environment as well as the challenges maritime nations face in the global common. Many nations have adopted the MDA framework in partial or in whole to build a comprehensive situational awareness to deal with growing threats in the maritime domain. India has also incorporated some ideas from the US MDA frameworks like situational awareness and strengthening information-sharing mechanism to expand India's maritime security role in the region. The Informational Fusion Centre of Indian Navy is a testimonial of India's vision to strengthen maritime security in the region and beyond. Similarly, the UDA also works on the same principle that information-sharing with the shareholder and the navy to create a common operating picture would significantly boost the navy's chance in mitigating challenges in the undersea areas.

The UDA is a subset of the MDA and shares the same basic tenet of the naval strategy – "before an enemy can be engaged, he must first be found". The naval strategy is mostly involved in detecting and tracking enemy movements and securing the maritime frontiers from the enemy's ships, submarines, and aircraft. During peacetime, the navy's primary role also involves safety and security of trade and energy sea routes. Maritime nations that immensely depend on the sea routes for energy and trade transportation realized the importance in building a MDA that integrates both civil and military aspects to protect global commons. The major challenge for the navy in performing these duties arises from the sheer vastness of the ocean, the turbulent seas/rough weather, and shipping traffic near crucial sea lanes. It makes the surveillance complicated if not difficult. The adversaries may try to take advantage of the situation to conduct military operations. Equally, the subsurface challenge is also drastically increasing with time and poses a major challenge to navies across the globe. Unlike surface targets, detecting and tracking underwater objects like submarines and drones are not just a difficult task but also an expensive one. It requires a systemic approach to handle the underwater challenges because the underwater monitoring or observing of the ocean has evolved into three-dimensional networks, including satellite remote sensing, land-based ocean observation stations, sea surface buoy arrays, scientific research vessels, submarine buoy arrays, underwater profile buoys, and sea bottom observation networks (Liu et al., 2019). Moreover, the maritime nations' growing interest in the seabed and Exclusive Economic Zones (EEZs) for resources and economic needs requires the need for building a comprehensive situational awareness, not only from a military and strategic perspective but also from an economic and development angle. That's why there is an urgent need to focus on the UDA.

Identifying the friend or foe is another great challenge in the underwater domain. Involvement of commercial and non-government entities in undersea activity has been gradually increasing over the period. The underwater provides a huge opportunity and avenue for researchers, industries, and environmental activists to understand the nature of the undersea marine environment. The use of underwater drones/gliders for commercial and scientific research has also increased in the recent past in the Indian Ocean. At the same time, there is apprehension about use of underwater vehicles for illegal activities like smuggling drugs, contraband, and terrorist activity. The Liberation Tiger of Tamil Eelam (LTTE) had used semi-submersible vehicles in their naval operations against the Sri Lankan Navy in the early 2000s. Even though the LTTE lost the war in 2009, terrorist groups or insurgent groups across the globe may try to replicate LTTE style of attacks on the naval warships, port, and other maritime infrastructure. The drug cartels in South America also use mini-submarines/semi-submersible vehicles to smuggle drugs illegally to the United States and Europe. In Spain, authorities have seized a semi-submersible vessel during a police raid on the mafia hideouts (Guy, 2021). The cross-Atlantic drug trade using semi-submersible vehicles clearly shows the sophistication of the drug cartels in designing and building such modern vessels for illegal trade. Therefore, it is going to be a difficult task for authorities in the future to differentiate between the commercial/economic/scientific activities from the malicious/illegal activity in the underwater domain. The Indian Ocean is already facing a such challenges from non-state criminal and illegal actors exploiting the gap in security system in the region. Regulating and governing the underwater domain is going to be tough challenge for India. Therefore, concepts like UDA are being increasingly discussed to create synergy between various authorities including defence, civil and corporate entity to mitigate such threats and address the safety and security of undersea environment.

India's Maritime Security Strategy: Maritime/ Underwater Domain Awareness

India conceptualized MDA based on its threat assessment in the region (Integrated Headquarters, Ministry of Defence, 2009). The major thrust area of Indian MDA includes surface, aerospace surveillance, and subsurface surveillance, effective communication, and information-based decision to discern between friend, foe, and neutral. The Indian Navy's strategic document, *Ensuring Secure Seas: Indian Maritime Security Strategy,* in 2015, has defined MDA as "an all-encompassing term that involves being cognizant of the position and intentions of all actors, whether own, hostile, or neutral, in all dimensions of a dynamic maritime environment, across the areas of interest" (Integrated Headquarters, Ministry of Defence, 2009). Building a comprehensive MDA is a continuous process and it will require constant gathering of intelligence, conduct of surveillance and reconnaissance of all dimensions (space, air, surface, underwater, and electronic), and data integration to derive actionable information for military operations. The Maritime Security Strategy 2015 noted the importance of UDA capability as a primary means for 'sea control'.

Characteristics of the Indian military maritime environment are a three-dimensional battle space in which naval forces must operate, viz., on, below, and above the sea surface (Annual report 2015–2016, 2016). The air, space, sea surface, and underwater affect every aspect of maritime warfare – surveillance, classification, localization, targeting, and weapon delivery, wherein the threat can come from any direction or dimension (Integrated Headquarters, Ministry of Defence, 2009). On the other hand, the biggest challenge in underwater surveillance is to detect and identify the threat in the region. Integrated Headquarters, Ministry of Defence (2009) has clearly described the challenges faced by India in underwater surveillances:

> The sea is very nearly opaque to a majority of sensors. In addition, the temperature and pressure profile and salinity conditions of the sea seriously impinge on the performance of underwater surveillance equipment. Submarines, for example, routinely use these conditions to their advantage, which are more pronounced in warm, saline waters.
>
> – Indian Maritime Doctrine, Indian Navy (2009)

Indian Maritime Security Strategy 2015 also suggests the need for the development of strong ASW capability to counter adversary submarine forces. This includes integral ASW capability within the naval fleet for its protection and for the deployment of dedicated forces in coastal waters and for open-ocean ASW (Ensuring Secure Seas: Indian Maritime Security Strategy, 2015). Essentially, the various geographic factors such as shallow water inhibit the acoustic environment, making underwater detection very challenging. Therefore, it is suggested that along with acoustic measures, the navy should also expand or develop off-board sensors that should be linked to a centralized system to get an underwater warfare picture and that, in turn, must be integrated into the overall command and control system to provide the Commander with a complete tactical picture.

Indian Defence agencies' earlier understanding on the underwater environment is limited to ASW and countermine operations; however, it has now additional responsibilities like protecting the sovereign resources from exploitation by corporate and external actors in the Indian EEZs. Indian marine scientific, environmental, and marine research agencies involved in monitoring the health of ocean floors have also inducted new technologies and sensors into the domain with capabilities often exceeding defence UDA sensors and system processing. Indian space programmes have developed some niche capability in ocean observation. In the last 30 years, the Indian Space Research Organisation (ISRO) has developed a series of application satellites relating to remote sensing, ocean observation, communication, and weather in orbit, and also has continuous plans for earth observation satellites providing optical and radar imaging. These data sets are the mainstay of operational oceanography. In situ observations, through tide gauges, moored buoys, ADCP moorings, drifting buoys, wave rider buoys, Argo floats, HF radars, etc., have provided observations to understand oceanic processes. Wave rider buoys provide information on wave climate in real time through the INSAT communication system. An indigenous drifter buoy has been developed having INSAT communication for measuring the sea surface temperature (SST), surface winds, and pressure. The large volume of data has been organized around GIS as the Ocean Data and Information System (ODIS). It provides data and information on the physical, chemical, optical surface, environment, and biological properties of oceans and coasts on various spatio-temporal domains that are vital for oceanographic research. Therefore, it is important to integrate various technologies and systems available in the country to build a comprehensive undersea picture.

India is scaling up its capability to detect, track, and identify moving underwater objects in the Indian Ocean. At the same time, India's UDA requirements are also fast changing with emerging non-military threats in the region. The search and rescue of sunken aircraft and vessels, blue economy, scientific exploration, and recreational activities are other factors that make the subsurface a crucial part of India's MDA/UDA. India is also the first respondent to any crisis in the region and has an important role to play. During the search of the missing Malaysian Airlines, MH-370 in 2014, as well as in the search of the missing Indonesian submarine in the Java Sea, the Indian Navy deployed its vessels and aircraft to assist in the search and rescue efforts. The Indian Navy also acquired two Deep Submergence Rescue Vehicles (DSRVs) to assist during the search and rescue of submarines in the IOR (Deep Submergence Rescue Vehicle Complex Inaugurated at Visakhapatnam, n.d.). The two DSRVs are currently located in the Visakhapatnam port in the east coast of India, which can be rapidly mobilized by air or road to facilitate submarine rescue operations even at distant locations. If India has to maintain its stature as a regional maritime power, it is important to achieve 'total awareness' of the situation in the Indian Ocean. This calls for constant monitoring and assessment of the condition of the oceans and the ocean's resources. An understanding of parameters impacting the processes and resources on and below the ocean surface is needed for effective UDA.

The UDA framework works on three levels:

1. Gathering data through monitoring the underwater environment for threats, activities, and resources, etc.
2. Using the generated data for planning security strategies, conservation plan, resource utilization, etc.
3. Formulating suitable strategy or regulatory framework and the monitoring mechanism at the regional and global levels.

This requires massive investment in building acoustic capabilities in the country to cover a particular region. It is a major challenge for developing countries like India, which is facing structural challenges in developing the UDA. The UDA is not a 'mere underwater extension of the MDA concept, but comprehensively addresses the safe, secure and sustainable growth model critically required in the Indo-Pacific strategic space'.

The UDA is an essential need in connection with ASW operations and manifests itself for effective monitoring of the ocean floor, sea surface, and subsurface. The extent to which UDA is exercised by different countries is diverse and dependent upon the country's maritime extent, its geopolitical standing, international cooperation, and financial conditions. In the Indian context, the UDA is a security architecture designed to detect and counter any subsurface security challenges in the region. Therefore, India is looking at various options including building a wide-area surveillance network in the Indian Ocean similar to that of the United States' SOSUS in the Atlantic Ocean. The Indian Navy would like to have ASW monitoring at important chokepoints on its eastern coast to detect and track submarines or other underwater objects entering the Indian waters. Indian Navy's ASW capability has to factor both manned (submarines) and unmanned underwater vehicles (UUVs) in the future operations. The inter-agency cooperation is also another important factor that should be taken into consideration while framing suitable UDA policy for India.

Bibliography

Aldinger, T., & Kao, J. (2005). Data Fusion and Theater Undersea Warfare – An Oceanographer's Perspective. In *Oceans '04 MTS/IEEE Techno-Ocean '04 (IEEE Cat. No.04CH37600)* (pp. 2008–2012). Kobe, Japan: IEEE.

Button, R. W., Kamp, J., Curtin, T. B., & Dryden, J. (2009). *A Survey of Missions for Unmanned Undersea Vehicles*. Santa Monica: RAND Corporation.

Clark, B. (2015). *The Emerging Era in Undersea Warfare*. Washington: The Center for Strategic and Budgetary Assessments.

Clark, B., Cropsey, S., & Walton, T. A. (2020). *Sustaining the Undersea Advantage: Transforming Anti-Submarine Warfare Using Autonomous Systems*. Washington: Hudson Institute.

DARPA Selects Performers to Advance Unmanned Underwater Vehicle Project. (2021, 2 May). *DARPA*. Retrieved from: www.darpa.mil/news-events/2021-02-05a.

Deep Submergence Rescue Vehicle Complex Inaugurated at Visakhapatnam. (n.d.). *Indian Navy*. Retrieved from: www.indiannavy.nic.in/node/26439 (Accessed on September 15, 2021).

Department of Homeland Security. (2005). *National Plan to Achieve Maritime Domain Awareness*. Washington: Department of Homeland Security.

Eckstein, M. (2019, November 8). Navy Undersea Warfare Priorities: Strategic Deterrence, Lethality and Networked Systems. *USNI News*. Retrieved from: https://news.usni.org/2019/11/08/navy-undersea-warfare-priorities-strategic-deterrence-lethality-and-networked-systems (Accessed on November 11, 2021).

Finch, D. (2011). Comprehensive Undersea Domain Awareness: A Conceptual Model. *Canadian Naval Review*, 21–26.

Friedman, N. (2019). *Strategic Submarines and Strategic Stability: Looking Towards The 2030s*. Canberra: Australian National University (ANU).

Goward, D. A. (2008). Maritime Domain Awareness: The Key to Maritime Security. In J. N. Moore, M. H. Nordquist, R. Wolfrum, & R. Long (eds.), *Legal Challenges in Maritime Security* (pp. 513–526). London: Brill.

Guy, J. (2021, March 15). First Spanish-Made 'Narco Sub' Found in Mediterranean Warehouse. *CNN*. Retrieved from: https://edition.cnn.com/2021/03/15/europe/drug-trafficking-semi-submersible-malaga-spain-scli-intl/index.html (Accessed on October 1, 2021).

Integrated Headquarters, Ministry of Defence. (2009). *Indian Maritime Doctrine 2009*. Sivakasi: Integrated Headquarters, Ministry of Defence (Navy).

Integrated Headquarters, Ministry of Defence. (2015). *Ensuring Secure Seas: Indian Maritime Security Strategy*. New Delhi: Integrated Headquarters, Ministry of Defence (Navy).

Liu, L., Liao, Z., Chen, C., Chen, J., Niu, J., Jia, Y., . . . Liu, T. (2019). A Seabed Real-Time Sensing System for In-Situ Long-Term Multi-Parameter Observation Applications. *Sensors*, 19.

Medcalf, R., Mansted, K., Frühling, S., & Goldrick, J. (2020). *The Future of the Undersea Deterrent: A Global Survey*. Canberra: Indo-Pacific Strategy Series, ANU.

Ministry of Defence Government of India. (2016). *Annual Report 2015–2016*. New Delhi: Ministry of Defence Government of India.

Toti, W. J. (2014, June). *The Hunt for Full-Spectrum ASW*. U.S. Naval Institute. Retrieved from: www.usni.org/magazines/proceedings/2014/june/hunt-full-spectrum-asw (Accessed on February 11, 2021).

United States Submarine Forces. (2012). *Design for Undersea Warfare Update One*. Norfolk, VA: United States Submarine Forces.

5 Assessing India's ASW Capability in the Indian Ocean

Given the overlapping maritime boundaries in South and Southeast Asia, given the presence and forays of navies outside the region into these waters, given the richness of the fishing zones and other blue economic indicators, and given the exploitable riches of the seabed in the EEZ and the difficult geopolitical scenario, situational awareness becomes very important for the actors in the region. For India, all these factors matter and some factors like geopolitics and neighbours assume greater importance and relevance. For India to safeguard its long maritime boundary and its EEZ under these developing circumstances, it becomes imperative to have maritime and underwater domain awareness.

The Indian Navy is an important naval force in the region and has played an active role in patrolling, anti-piracy operations, HADR, and joint naval exercises with a number of participating countries. Though its resources and capabilities are substantial, India needs to increase its capabilities to fulfil its safety and security needs. These capabilities include augmentation in the underwater, surface, and air/space domains. Logically, Anti-Submarine Warfare capability enhancement occupies a top slot in planning and the Indian Navy addresses it under strategic, operational, and tactical categories.

At the **strategic level**, ASW is aimed at projecting power in the maritime environment and safeguarding its **Area of Interest**. The intent of the power projection is to deter enemy submarines from venturing into the Area of Interest and to demonstrate the capability to contain, limit, or even destroy the warfighting capacity of the intruder.

At the **operational level**, the primary focus is to determine where and when to defeat the enemy submarines.

At the **tactical level**, ASW function looks at the local condition of the platforms, weapons, and sensors in the area of encounter. The role is to detect, track, and classify all intruding submarine activity.

DOI: 10.4324/9781003298380-5

The Indian Navy's Maritime Security Strategy, 2015, provides some insight into the Indian Navy's ASW strategy. The primary focus of ASW is based on the 'Shore-based/Airborne ASW and ASW Ships & Submarine'. The air-based and ship/submarine-based ASW operates in conjunction with each other as part of a coordinated ASW effort (Integrated Headquarters, Ministry of Defence, 2015). A detailed assessment on the subject provides an overview of the Indian Navy's present capability in countering the Chinese threat in the Indian Ocean.

Indian Navy's Shore-Based/Airborne Anti-Submarine Warfare

The navy's ability to carry out the tasks mentioned earlier critically depends on the available technology. The radar and sensors, including electromagnetic sensors, are the 'eyes and ears' of the naval force in detecting and tracking the enemy submarines. The Indian Navy uses ship-based sonar systems to detect, track, and classify the submarines/underwater objects in its daily routine operations. India is also currently scouting for new emerging technologies like unmanned systems, including data analytics and artificial intelligence, and underwater drones/gliders, which can provide greater depth in understanding the multi-domain Common Operational Picture (COP) and which are increasingly playing an important role in the ASW.

In modern naval warfare, naval aviation plays an important role in monitoring and providing real-time information of the Area of Interest to the fleet commander. This real-time information is crucial in maritime reconnaissance, it can be gathered through various platforms including space-based systems, long-range maritime patrol aircraft, ship-borne systems, airborne/shore-based early warning radars, and UAVs. Since the late 1980s, the Indian Navy has used platforms, such as IL-38SD and Tu-142M (acquired from Russia), for carrying out long-range maritime patrol/surveillance missions in the Indian Ocean. These platforms have an operating range upto the order of 6,000+ km and are equipped to carry out surveillance, search, and rescue; maritime reconnaissance; and ASW operations (De-induction of Indian Navy's TU142M Aircraft and Induction of Boeing P8 I into INAS 312, n.d.). In addition, IL38 can detect and intercept surface vessels and submarines. During Operation Cactus in 1988, TU-142M Maritime Reconnaissance and Airborne Anti-Submarine Warfare aircraft played a crucial role in detecting and tracking fleeing mercenaries from Maldives until they were apprehended by an Indian warship (After 29 years of service, Navy bids adieu to TU-142M, 2017). The Russian TU-142M and the IL-38SD patrol crucial chokepoints from the Strait of Malacca to the Strait of Hormuz for any possible surface and subsurface threats from both state

Table 5.1 List of ASW Squadron and Dornier Squadron

	Squadron name	Base/location	Aircraft name	Role
1	INAS 310	INS Hansa/Goa	Dornier 228	Maritime Patrol
2	INAS 311	INS Dega/Visakhapatnam	Dornier 228	Maritime Patrol
3	INAS 312	INS Rajali/Arakonnam	Boeing P-8I Poseidon	Long-Range Maritime Patrol/ASW
4	INAS 313	Chennai	Dornier 228	Maritime Patrol
5	INAS 314	Porbandar	Dornier 228	Maritime Patrol
6	INAS 315	INS Hansa/Goa	Ilyushin IL-38	Long-Range Maritime Patrol/ASW
7	INAS 318	INS Utkrosh/Port Blair	Dornier 228	Maritime Patrol
8	INAS 330	INS Shikra/Mumbai	Sea King 42B	ASW
9	INAS 333	INS Dega/Visakhapatnam	Kamov Ka-28	ASW
10	INAS 336	INS Garuda/Kochi	Sea King 42B	ASW Training
11	INAS 350	INS Dega/Visakhapatnam	Sikorsky UH-3H	ASW
12	INAS 550	INS Venduruthy/Kochi	Dornier 228	Maritime Patrol

Source: Compiled by author from the Indian Navy website.

Table 5.2 Indian Procurement of ASW Aircraft from the United States and Russia, 2000–2019

Year	Country	System	Nos.	Value in US dollars
2001	Russia	IL-38SD – ASW	5	150 million
2009	USA	P-8I Neptune – ASW	8	2.1 billion
2016	USA	P-8I Neptune – ASW	4	1 billion
2019	USA	P-8I Neptune – ASW	6	3.2 billion

Source: Compiled by author.

and non-state actors. The aircraft also played a significant role in Operation Vijay (1998), Operation Parakram (2002), and anti-piracy operations off the coast of Somalia from 2011 onwards. In the wake of the growing underwater threat in the region, the Indian Navy has decided to expand its fleet of long-range maritime surveillance aircraft. In 2013, the Indian Navy inducted a new batch of Boeing P-8I aircraft. The IL-38SD underwent a mid-life upgradation and continues to serve in ASW missions. The Dornier 228 also played a key role in the 1965 and 1971 War with Pakistan. The Dornier 228 aircraft is also used in Anti-Submarine Warfare in the coastal waters to detect the submarines or other illegal activities at sea.

Table 5.3 Long-Range Maritime Patrol and ASW Aircraft

	IL-38SD (Sea Dragon)	*Boeing P-8I*
Role	• ASW & ASV mission • SAR • Limited Intensity Maritime Operation	• ASW & ASV mission • SAR • Offensive operation through Air-Launch Missile.
Capability	• Accurate Navigation • Ability track multiple targets • Positive identification of targets • Surface plot transfer • Computerized weapon release • Over the Horizon Targeting • Strike homing • Air-sub cooperation	• Sustain without refuelling in HI-LO-HI missions. • Interoperable with manned/ unmanned platforms. • Can fly both subsonic speeds and at slow patrol speed. • Fuel efficiency even at low speed.
Equipment	• Radar (SD1) • Radio Sonic System (SD2) • Magnetometric System (SD3) • Control Computer System (SD4) • Infrared Television System (SD5) • SD6 ESM System and Radar Fingerprinting System • Trigun (AIS-MDA-Hurricane)	• The Maritime Patrol Radar (MPR) • Sonic System • CAE AN/ASQ-508A Magnetic Anomaly Detector (MAD) • Electromagnetic Support Measure (ESM) • Electro-Optical Device/ Forward Looking Infrared • Missile Approach Warning Systems (MAWS) • Integrated Data Link • Armament Package

Source: Compiled by author based on the inputs from (Sharma, 2013), (21st Century Maritime Security for the Indian Navy, n.d.).

The IL-38SD and Boeing P-8I brings hitherto unparalleled capabilities to the Indian Navy. But, MoD and the Indian Navy's emphasis on procuring more Boeing P-8I for the Indian Navy clearly shows that it has an important role to play in India's maritime security. The first eight batches of P-8I aircraft were inducted into the Indian Navy in 2015; it was deployed in INS Rajali. In 2016, the government approved to buy additional four Boeing P-8I aircraft, which are scheduled for delivery in 2021 (Reim, 2021). In addition, in 2019, the Defence Acquisition Committee (DAC) gave approval to procure six more Boeing P-8I aircraft for the Indian Navy (Defence Acquisition Council, chaired by Raksha Mantri Shri Rajnath Singh, approves Capital Procurement for the Services amounting to over Rs 22,800 crore, 2019). So, by the end of 2030 the Indian Navy will be operating close to 18

P-8I aircraft for ASW and other maritime security operations in the Indian Ocean.

According to the Indian Navy, the P-8I aircraft is equipped for long-range anti-submarine warfare, anti-surface warfare, intelligence, surveillance, and reconnaissance in support of the broad area, maritime, and littoral operations. Boeing P-8I has the strength and capability to operate in low altitude. It is designed to perform most of the operations from high altitudes, where the thinner atmosphere allows for greater fuel efficiency and gives better vantage for some of the sensors. The P-8I can loiter overhead as its lowest speed is 200 miles per hour, and it can stay on the mission for an extended period due to its in-flight refuelling capabilities. The P-8I's primary payload is a diverse ray of sensors that includes two radar sets – a forwarded-mounted Raytheon AN/APY-10 and an aft-mounted Telephonic APS-143C(V)3 for 360-degree coverage (Mazumdar, 2013). The MX-20 EO/IR systems, which contain advanced multi-sensor imaging and laser sensors, creates accurate situational awareness pictures (WESCAM MX™-15, Air Surveillance and Reconnaissance, n.d.).

Number of key systems in the aircraft designed to track submerged submarines, the rotary launcher system in the rear of the P-8I can dispense sonobouys into the water. The P-8I has its own acoustic sensor and even the new hydrocarbon sensor that can sniff fuel-vapour from submarines. The P-8I computers are designed to fuse data into a single coherent picture for the operators and share the data through a secured link to the naval warships and airplanes. The data link and communication onboard of P-8I including SATCOM and Link II tactical datalinks were developed indigenously in India (Indian-designed Data Link II Delivered to Boeing, 2010). Apart from the sensors and electronic warfare systems, it is also capable of carrying a huge range of weapons payload depending on the flight mission. The US Navy operates the most advanced version of this aircraft and is equipped with the newer MAC active sonobuoy systems, which can achieve 5- to 8-nm detection ranges and, therefore, requires half as many sonobuoys to cover the same area (Clark, 2015).

The long-range maritime reconnaissance aircraft is an integral part of the Indian Navy's ASW capability to detect and track Chinese submarines far from the Indian coast. The patrolling areas of interest of these aircraft are the Straits of Malacca, Lombok, and Sunda, where Chinese nuclear and diesel submarines are frequently spotted. India's strategic partners, the United States, and Japan also operate a similar variant of the aircraft that helps the Indian Navy to train and learn from other navies to effectively use the aircraft in submarine warfare. Information-sharing becomes a lot easier and convenient as they share similar systems. The JMSDF has a vast experience in operating P-3 Orion/P-8 Poseidon over many decades in the

Western Pacific and the Sea of Japan to track Russian and Chinese submarine operations. Training with advanced navies would enable Indian Navy operation capabilities in the Indian Ocean Region; at the same time, it also complements the multilateral security cooperation initiative to maintain 'good order at sea'.

Medium-Range Maritime Reconnaissance Aircraft and Anti-Submarine Warfare Helicopter

India has 7,500-km-long coastlines and its Exclusive Economic Zone (EEZ) is approximately around 77,000 square miles, including 1,208 uninhabited islands, which pose a significant challenge to the Indian Navy in monitoring and surveillance of the vast maritime space. The navy has six squadrons of Dornier 228 aircraft, helicopters, and UAVs for the maritime reconnaissance and surveillance of the coastlines and the EEZ. Indian Navy operates six squadrons of Dornier 228–201s, which are built locally by HAL and fitted with wider varieties of Intelligence, Surveillance, and Reconnaissance (ISR) sensors, payloads like Elta's EL/M 2022 V3 multimode radar, CoMPASS electro-optical (EO) pod, ESM, and Trigun AIS to provide real-time information to naval commanders, as well as to spot periscopes of enemy submarines lurking in the coastal waters. Dornier aircraft are the workhorse of the Indian Navy and their primary role is monitoring coastal waters and EEZs for possible threats from enemy submarines sailing into the Indian waters. During war time, Indian Coast Guard Dornier aircrafts can instantaneously switch over to ASW role. Both Indian naval and Coast Guard personnel are trained in an 'Airborne Tactics' course for 36 weeks, in various subjects including tactics employed in air warfare, anti-submarine warfare, and exploitation of airborne avionic systems (Airborne Tacticians for Navy and Coast Guard, n.d.). The Indian Navy also has an Anti-Submarine Warfare School in the Southern Naval Command at Kochi, which trains Indian naval officers and sailors in Anti-Submarine Warfare. The ASW School is equipped with a state-of-the-art simulator, which is capable of simulating acoustic signals encountered at sea during ASW operations (Commissioning of Advanced LOFAR Simulator at ASW School, n.d.). In 2019, the ASW School inducted an Advanced Low Frequency and Ranging (LOFAR) simulator to train LOFAR operators in the analysis and classification of acoustic signatures of the target by using audio/video techniques (Commissioning of Advanced LOFAR Simulator at ASW School, n.d.). The DRDO's Naval Physical and Oceanographic Laboratory (NPOL) conducting research in undersea warfare is located in Kochi and works closely with the Indian Navy's ASW School in training officers in submarine warfare.

The search helicopter fitted with sonobuoy or dunking sonar would be dispatched to detect, track, and destroy enemy submarines. The deployment of sonobuoy or dunking sonar is based on the type of mission the search helicopter is designated to carry out. The four primary differences between dunking sonar and sonobuoy are (Yoash, 2016) as follows:

1. Deploying dunking sonar takes time; it takes less time to deploy sonobuoy.
2. The detection radius is larger in dunking sonar.
3. Once the sonobuoy is dropped, it monitors the area continuously till it dies out.
4. Helicopters can carry out only a limited number of sonobuoy due to space constraints.

To bridge the difference, at times both sonobuoy and dunking sonar were used to achieve greater coverage. The passive sonobuoy can actively monitor the underwater sound and transmit the data to a helicopter within the range for the fixed period and also supplement the helicopter with dunking sonar to find and track the submarine underwater. However, both are vulnerable to difficult sonar conditions like shallow waters, multiple layering, sharp thermoclines, reverberation, and potentially high noise levels, all of which complicate the detection of small targets (Martin, Ogle, Whalen, & Wignall, 2006). Experts claim that for protracted naval operations, the sensor of choice would be the dunking sonar (Martin, Ogle, Whalen, & Wignall, 2006). The probability of finding the submarine is also dependent upon the deployment of search helicopters.

The helicopters are dispatched on ASW missions based on inputs from the maritime intelligence about the location (range and direction) of a submarine target. The point of detection is called a *datum*. According to naval experts, there are three ways to improve detection of the enemy submarine by using search helicopters (Yoash, 2016):

1. Reaching the *datum* on time, dunking sonar detection radius, and the time it takes to dip the sonar are the crucial factors that help the search helicopter to detect the submarine in normal climatic conditions.
2. The speed of the helicopter to fly to the *datum* increases the probability of detecting the submarines.
3. Employing two or more helicopters to the *datum* significantly increases the probability of detecting the submarine. Importantly, the initial cues (in the form of submarine bearings or speed) from intelligence agencies, from the Indian Navy's P-8I patrolling the operational area, the Indian Navy Task Force patrolling the operational area, or inputs from friendly naval ships and aircraft are crucial for detecting a submarine.

The Indian Navy currently operates three types of helicopters for ASW purposes – Sea King 42, Sikorsky UH 3H, and Kamov-28. These helicopters bring different capabilities to the Indian Navy's ASW operations. These helicopters are equipped with state-of-the-art maritime radars optimized to detect submarine snorkels or periscopes. Integrated ESM technology, mounted MAD, dipping sonars, and dropped active/passive sonobuoys are used to detect and track enemy submarines, while data link capabilities provide real-time information to the naval commander. The ASW aircraft and helicopters possess advanced technologies and architecture that can be modernized and be fully networked with acoustic and non-acoustic sensors to take advantage of the maritime space. The new, modern technologies and advanced computing provide MPA and helicopters a crucial advantage to identify, track, and classify the target even in difficult sonar conditions. Given its proficiency to track underwater targets even in difficult sonar conditions, the airborne ASW systems have now become preferred options for navies worldwide. For effective coverage of the area, the Indian Navy has inked an agreement with Lockheed Martin to procure 24 Sikorsky MH60-R Multi-Role helicopters, through a government-to-government deal. The first batch of two MH60-R Multi-Role helicopters was delivered to the Indian Navy (PTI, 2021). In the next few years, Indian Navy's modern destroyers and frigates will be inducted with the modern MH60-R helicopters for ASW operations. Airpower is an important aspect of maritime reconnaissance, which provides basic inputs for a common operational picture. India's acquisition of P-8I long-range maritime patrol aircraft, along with the new MH60-R helicopters, would significantly boost its ASW surveillance and engagement capabilities.

Surface Ships and Submarines

The Indian Navy's surface combatants and submarine force is equipped to counter any submarine threat. *The Delhi-, Kolkata-, Shivalik-,* and *Rajput-*class destroyers together with the *Talwar-, Shivalik-,* and *Brahmaputra-*class frigates and the newly commissioned *Kamorta-*class ASW corvettes give the navy a formidable ASW capability. The destroyers and frigates have been equipped with towed array sonar to hull-mounted sonar to detect, track, and classify any underwater target in the AOI. In addition to that, the Indian Navy is also planning to induct 16 ASW shallow water crafts equipped with a hull-mounted sonar and a Low Frequency Variable-Depth Sonar (LFVDS) (GRSE starts production of anti-submarine ship, 2020). The shallow water crafts would be a great addition to the Indian Navy to maintain their presence in the coastal waters against the Chinese and Pakistani submarines or underwater drones. Pakistan operates both diesel and

Table 5.4 Indian Navy's Principal Surface and Subsurface Platforms for ASW Operations

Surface combatants	
Destroyers	*Delhi-class, Kolkata-class, Shivalik-class, Rajput-class*
Frigates	*Brahmaputra-class, Godavari-class, Talwar-class, Kamorta-class*
Submarines	
SSBN	*Arihant-class*
SSN	INS *Chakra* (Leased from Russia)
SSK	*Shishumar-class, Sindhughosh-class, Kalvari-class*

Source: Compiled by author.

midget submarines capable of reaching the Indian west coast. Tracking these submarines in the busy sea lanes of the Arabian Sea is a major underwater challenge for India.

The Indian Navy submarine programme is aimed at building a credible maritime force in the Indian Ocean Region. The Indian Navy is planning to acquire 24 new submarines, including six nuclear attack submarines to narrow the growing gap with the growing underwater competition in the region (Defence Ministry issues formal tender for mega submarine programme, 2021). The Arihant-Class and the INS Chakra are the two deployed nuclear submarines. The Indian Navy is planning to induct newer SSBNs and SSNs to maintain a sea-based deterrence capability as part of its 'nuclear triad'. India is also likely to commission second Arihant-class submarine *Arighat*, which was launched in 2017 and is currently in the final stages of sea trails (Bedi, 2020). The DRDO is also planning to launch new *Arighat* successor SSBNs designated S4 and S4 (Star), displacing over 1,000 tonnes more than Arihant-Class, which is currently undergoing construction at the Ship Building Centre (SBC) in Visakhapatnam (Unnithan, 2017). As INS Chakra is nearing its completion of leasing period, India has signed a 10-year lease agreement with Russia for new SSN, which will join the Indian fleet around 2025 onwards to augment India's underwater domain awarness (Shukla, 2019).

Diesel submarines account for a huge share and play an important role in intelligence, surveillance, and reconnaissance against enemy activity in the Indian Ocean. The *Sindhughosh-class and Shishumar-class* constitute the major share of India's submarine strength. But, Indian Navy has also planned to include a new diesel-electric submarine like *Kalveri*-class submarines with advanced acoustic absorption techniques, low radiated noise levels, and hydro-dynamically optimized shape to maintain stealth and to maintain tactical control. The *Kalvari-class* submarines are the backbone of India's conventional attack submarines. So far, four *Kalveri*-class submarines have been inducted into the service with advanced air-independent propulsion (AIP) system to increase endurance at sea. DRDO has in collaboration with Larsen & Toubro, Thermax, and the Centre for Development of Advanced Computing (C-DAC) indigenously developed AIP system based on phosphoric acid fuel cell (PAFC) technology (Scott, 2021). The new AIP system will be installed in all the six *Kalvari-class* submarines as a retrofit during their first major overhaul (Scott, 2021). The *Kalvari-class* is also equipped with 'S-Cube Integrated and Modular Submarine Sonar Suite', which can provide cutting-edge passive and active performance in deep or littoral waters to detect, classify, locate, and track a full range of threats (S-Cube Integrated Submarine Sonar Suite, n.d.). Apart from the weapon and sensor platforms, the navy's deployment/operations in the Indian Ocean is a key factor to ensure the safety of the nation. The deployment of vessels in crucial waterways, chokepoints, and ports in the region increase the chances of spotting/monitoring Chinese submarine movement in the region. The navy ships and helicopters deployed in long patrol would increase the Indian Navy's chance in gathering information or intelligence against the Chinese and Pakistani naval patrol in the IOR.

Indian Navy Deployment in Anti-Submarine Warfare Operations

The biggest challenge for the navy in combating subsurface threats emerges far away from the Indian coast. This requires constant monitoring of chokepoints, high seas, and important sea lanes against subsurface threats. To optimize the Indian Navy's pursuit, naval ships have been regularly deployed for Presence and Surveillance Missions (PMS), off-critical chokepoints and international sea lanes in IOR. Under the Mission-Based Deployment (MBD), 15 naval warships patrol 7 areas, and the areas include (New Mission-Based Deployment concept to result in greater presence and visibility in IOR – Admiral Sunil Lanba, 2017)

1. MALDEP (Strait of Malacca),
2. NORDEP (North of the Bay of Bengal),
3. ANDEP (North Andaman and South Nicobar),
4. GULFDEP (North Arabian Sea),
5. POGDEP (Anti-piracy in Gulf of Aden),
6. CENDEP (Off Maldives and Sri Lanka), and
7. IODEP (Off Mauritius and Seychelles).

The primary mission is to check all the entry and exit routes to and from the Indian Ocean. Each warship is turned around after 3 months of deployment.

Admiral Sunil Lamba, former CNS, introduced the MBD and enhanced the Indian naval unit to be a first responder in the region in case of any crisis and showcase the Indian Navy as a 'Net Security Provider' in the region. The Presence and Surveillance Mission is also critical for the Indian Navy to gather active intelligence on littoral states and Chinese naval deployment in the region.

The deploying mission-based ships and aircraft in the Indian Ocean also require sustained effort from India to maintain logistics, spares management, forecast refit, and expenditure management. The Indian Navy has planned to mitigate the challenges through cooperation initiatives with friendly countries in the region. So far, India has signed four military logistic agreements with the United States, Singapore, France, Japan, and South Korea. Similar agreements are in the pipeline to be signed with Russia, the UK, and Australia. This will ease India's logistic problem and it will meet the navy's requirements to maintain round-the-clock watch in the Indian Ocean (Thomas, 2019). Since signing of the Logistic Exchange of Memorandum of Agreement (LEMOA) with the United States, in 2016, the Indian naval ships patrolling in the Gulf of Aden have been refuelled by the US Navy tankers; similarly, the Indian naval ships and aircraft are using the French naval base in Djibouti for a quick turnaround. The agreement partner countries provide the Indian naval ships and long-range maritime patrol aircraft to use their base for landing and refuelling of aircraft, thereby extending the operational envelope to a substantial degree (Thomas, 2019).

The chokepoints in the IOR are the main entry points for foreign submarines. Indian Navy has organized its assets in a pattern such that it can detect the enemy submarine as soon as it enters the geographical chokepoints in the region. Preliminary research shows that the Indian Navy's ASW strategy is somewhat similar to the United States' Cold War ASW tactics called the 'Gauntlet Pattern', where submarines would be driven within the barriers organized along the geographical chokepoints to major seas and/or at exits from key Soviet bases. Then, a ship-based helicopter or land-based aircraft P-3 would conduct sweeps in areas known for their concentration of submarines, and utilize sonobuoys to detect and localize them (Zakheim, 1977). Lastly, surface ships or submarines have been engaged in the pursuit of forcing the enemy submarine to the surface or destroying the submarine using depth charges or ASW missiles/rockets.

The Indian Navy is facing two sets of challenges in assessing the security threat in the region. The first one is a lack of resources and preparation to monitor the chokepoints (with seabed sensors or array of sensors) or gather information of the entry and exits of submarines or subsurface threats. Second, re-acquiring or tracking of submarines in the high seas remains a significant challenge, the seabed sensors or sonobuoy alone may not be adequate to provide truly robust anti-submarine surveillance. Therefore, the Indian Navy should create an ASW architecture to bridge the gap in surveillance near the chokepoints. Once the seabed sensors are deployed at chokepoints, then the Indian Navy can augment satellite-based wide-area surveillance or surface/submarine/unmanned sensors for initial detections (Kapur & McQuilkin, 2017). These detections would be handed off to fixed- or rotary-wing air or surface platforms to maintain a track of underwater objects. Effective integration of these data and information provides an operational picture for the naval commander to assess the seriousness of the threat. The information from Electronic Intelligence (ELINT) and Communication Intelligence (COMINT) can provide additional information in detecting subsurface threats. A combination of air, surface, and submarine ASW capabilities will help India to identify the logical path forward in a resource-scarce environment.

Building Underwater Domain Awareness: Assessing DRDO Capability

Indian observation of the Indian Ocean started only a couple of decades ago. So, the region is inadequately observed, and this is a work in progress. The initial observations are limited to military use, like collecting intelligence, surveillance, and information suitable for anti-submarine warfare (ASW) and anti-mine warfare. The Indian Ocean environment has a large effect on acoustic propagation; therefore, studies on the sonar performance, acoustic methods, and mapping the underwater landmass using acoustic methods have been developed in India. The Indian research institutes, including ISRO, IITs, CSIR Lab, and DRDO Labs – NPOL and Marine Research Lab, have led to the development of various applications for military use. In the wake of the growing importance of underwater domain awareness, monitoring sea surface, and underwater for any threats, through the deployment of acoustics, non-acoustics, and space assets becomes an inescapable requirement for India. These steps are essential to maintain a strategic situational awareness to effectively implement 'sea-control' or 'sea-denial' in the Indian Ocean. A systematic study of acoustic propagation in the underwater realm has led to remarkable advances in sonar technology. DRDO's naval vertical is pioneering research in sonar technologies in India.

DRDO's technology focus is primarily on developing underwater surveillance systems for the Indian Navy through indigenous efforts. DRDO is pursuing research and development in advanced sonar transducer technology, sonar power amplifiers, sonar signal processing, information and display processing capabilities; moreover, sonar deployment technology integrated with new materials for vibration isolation has taken a leap

Table 5.5 Underwater Surveillance System Developed by DRDO

Technologies	Lab	Features
ABHAY is compact hull-mounted sonar System for ships	NPOL	Passive and active sonar system designed for corvettes and patrol vessels for ASW
Advanced Panoramic Sonar Hull Mounted (APSOH)	NPOL	First sonar developed by NPOL in 1983
Hull-Mounted Sonar Advanced (HUMSA)	NPOL	Second generation of sonar where the sonar array depth can be varied for tactical purposes
HUMSA-NG	NPOL	Third-generation active and passive sonar, hull mounted, can be fitted in frigates, destroyers, ASW corvettes, and other classes of ships
Compact Hull-Mounted Sonar System for Ships (HMS-X2)	Naval systems and Materials (NS&M)	Compact active cum passive sonar system, developed for Anti-Submarine Warfare (AWS), corvettes, shallow water crafts, and patrol vessels
HUMSA-UG	NPOL	HUMSA-UG is an advanced integrated active cum passive sonar system designed for ASW platforms.
MIHIR-Mid Frequency Airborne Sonar	NPOL	First-generation helicopter sonar system, comprising dunking sonar and a four-channel sonobuoy processor
Low Frequency Dunking Sonar (LFDS)	NPOL	An advanced version of MIHIR, LFDS is capable of both active and passive sonar operations
Nagan – Towed Array Sonar	NPOL	Long passive array towed behind the ship, towed array can be deployed up to a depth of 200 m
PANCHENDRIYA – submarine Sonar	NPOL	The sonar consists of passive surveillance, passive ranging, intercept, active, and underwater communication systems
Payal – submarine Sonar Suite	NPOL	Passive, active, intercept, and underwater communication systems
USHUS – submarine Sonar Suite	NPOL	The suite includes passive sonar, active sonar, intercept sonar, obstacle avoidance sonar, and underwater telephony
USHUS-2 – Submarine Sonar Suite	NPOL	USHUS-2 is a world-class sonar suite, tailored for the remaining four EKM-class of submarines

(Continued)

Table 5.5 (Continued)

Technologies	Lab	Features
Tadpole – sonobuoy	NPOL	Deployable sonobuoy from high altitudes, two selectable operating depths, and three different operating periods for up to 8 hours
Underwater Wireless Acoustic Communication System (UWACS TRITON)	NPOL	Built on Software Defined Radio (SDR) architecture
Maareech – Advanced Torpedo Defence System (ATDS)	NPOL	Torpedo detection and countermeasure capability
Advanced Light Towed Array Sonar (ALTAS)	NPOL	Capable of detecting, locating, and classification of submarines operating below the sonic layer
Seabed Array	NPOL	Sonar meant for passive surveillance of underwater targets

Source: (DRDO, 2017).

forward in realizing products for end applications (DRDO, 2017). These sonar systems are developed to meet user's requirements. These systems are installed and operated onboard the Indian naval surface ship, submarine, and airborne platforms.

Operational Status: In the 1980s, DRDO developed sonar systems for the Indian surface vessels and submarines built indigenously. At this time, most of the Indian naval vessels were using the Russian or western sonar systems for ASW operations. The Advanced Panoramic Sonar Hull Mounted (APSOH), a medium-frequency active/passive sonar, was the first sonar built by NPOL that was inducted into the Indian Navy in 1983 (DRDO, APSOH, n.d.a). The success of APSOH prompted DRDO to build an advanced version of APOSH system called Hull-Mounted Sonar Advanced (HUMSA) to deal with emerging threats from Pakistan's new submarine with AIP systems and China's emerging submarine capabilities in the 1990s. The HUMSA-NG (next generation) is an upgraded version of HUMSA with new receiver electronics and ultra-cool power amplifier systems installed and operated in the frontline ships of the Indian Navy. The Indian submarine programme has also received indigenously developed USHUS – submarine sonar. The USHUS sonar is the first indigenously developed integrated sonar system for the Indian submarine that has reached full operational status. The lesson learnt from the design and

Table 5.6 DRDO's Sonar Systems in Service with Indian Navy

Item/variant	Indian naval Platforms	Status
HUMSA, HUMSA-NG, HUMSA-UG	Brahmaputra (FFGHM); Delhi (DDGHM); Kamorta (FFGH); Kolkata (DDGHM); Shivalik (FFGHM); Talwar (FFGHM); Godavari (FFGHM); Rajput (Kashin II) (DDGHM)	In service
USHUS, USHUS – 2	Sindhughosh (SSK); (Kilo) (Project 877EKM); Arihant ((SSBN/SSGN)	In service
Maareech – Advanced Torpedo Defence System (ATDS)	Kolkata-class	In service

Source: Compiled by author.

development of PANCHENDRIYA sonar helped DRDO to develop more advanced USHUS sonar systems for the EKM-class of submarines and replace the Russian MGK-400 and MG-519 systems. The USHUS-2 is also getting installed in other indigenously designed and developed submarine platforms (DRDO, USHUS – 2, n.d.b). Maareech Advanced Torpedo Defence System (ATDS) is an indigenously developed anti-torpedo system jointly developed by NPOL and NSTL.

According to news reports, two production-grade systems manufactured by BEL have been installed and their user evaluation trail was completed onboard INS Gomati and INS Ganga (BEL inaugurates upgraded ATDS Maareech manufacturing facility, 2020). The BEL has launched a fully indigenized Maareech Integration Facility with a capacity to manufacture and deliver 12 systems per year (PIB, 2020). In other areas like dunking sonar for light helicopters and air launch sonar, NPOL and NS&M were able to meet the technology requirements, but the naval systems failed to meet the Indian Navy's operational requirements. Since, the Nagan – a towed array sonar – failed to meet Indian Navy requirements, the navy has placed an order to import six of the Atlas Elektronik's Active Towed Array Sonar (ACTAS) systems, which is a low-frequency ASW sonar system that operates simultaneously in active and passive modes and provides high-resolution target detection (Shukla, Sonar contract provides major boost to navy, 2014). According to the news reports, ACTAS were imported for Kamorta-class of ASW corvettes, but three have been fitted on Talwar-class frigates and three on Delhi-class destroyers (Chand, 2016).

For wide-area surveillance, the DRDO has developed a Towed Array Data Sonar System (to detect submarines and torpedoes). It has also developed a miniaturized underwater acoustic sensor called hydrophone; this hydrophone is based on the metal-oxide-semiconductor field-effect transistor (MOSFET) and piezo-electric sensors (Subramanian, 2012). These miniature sensors find applications in thin-lined towed arrays for submarines, ships, and unmanned surface vehicles. DRDO has also developed a conductivity, temperature, and depth sensor to measure the temperature, salinity, and depth of the ocean because the speed of sound in the ocean depends on these parameters and they influence the sonars' performance (Subramanian, 2012). These technologies demonstrate that the DRDO has capability in building underwater sensors and sonar for wide-area surveillance.

The NPOL plans to concentrate on an ASW suite for ships and submarines and to build a comprehensive wide-area surveillance network in the region. Based on the requirements from the Indian Navy, NPOL is in its initial stage of product design of two systems: INDUS-NET (Indian Distributed Underwater Surveillance Network) and Varuna MAALAA to measure ambient, acoustic-level analysis in the IOR.

The DRDO also successfully launched SAGAR MAITRI in alignment with India's policy or doctrine of maritime cooperation in the IOR – 'Safety And Growth of All in the Region (SAGAR)'. The aim is to promote closer cooperation in socio-economic aspects as well as greater scientific interaction, especially in ocean research among Indian Ocean Region (IOR) countries. The specific scientific component of 'MAITRI' (Marine & Allied Interdisciplinary Training and Research Initiative) sets INS Sagardhwani apart from the traditional oceanographic research vessels and helps it to take a lead role to pursue exploration of the deep sea in innovative ways. The objective of the Sagar Maitri programme are twofold: (1) Data collection from the entire north Indian Ocean with immediate focus on the Andaman Sea and the Malacca Strait for strategic application by deploying INS Sagardhwani and (2) Establishing long-term collaboration with eight IOR countries in the field of 'Ocean Research & Development'. This programme also aims to establish long-term scientific collaboration with IOR countries in the field of ocean research and development, and data collection with a focus on the Andaman Sea.

The DRDO ecosystem is well suited to develop a comprehensive UDA system for India. However, the budgetary constraints, lack of adequately trained manpower in the specialized field of sensors, sonar, and oceanographic research, and access to crucial technology from abroad are some of the few factors that hinder DRDO in developing breakthrough technology in underwater systems development. The new emerging technologies like big data, artificial intelligence, and additive manufacturing need to be integrated into the DRDO ecosystem to expedite the processes and minimize cost.

The lifespan of sonar is limited to 10–12 years and the oceanic properties are constantly changing due to climatic, biological, and anthropological activities. Therefore, detecting submarines in such difficult sonar conditions is an extremely difficult task. The proliferation of submarines with advanced AIP and nuclear technology increases the stealthy features of submarines in the high seas and littorals. Underwater acoustic systems, both active and passive, greatly contribute to fill this gap by listening to the ocean noise or by transmitting pulses and interpreting the received echoes to detect submarines or mines in open oceans. Now, the ASW sonar technology has grown exponentially in some of the advanced countries. Compared with western countries, India's ASW capability is limited and would require robust detection capability to maintain effective UDA over the Indian Ocean.

The wide-area surveillance in the Indian Ocean is only possible by building a thorough domestic, technological, and enhanced acoustic capability. This requires a series of research studies to understand the various aspects of the underwater environment in the Indian Ocean Region. Gathering underwater data collected across the IOR would be the first step to understand the diurnal and seasonal conditions. India's scientific research in the field of physical oceanography, chemical oceanography, marine ecology, and marine geology to understand the oceanic condition has not progressed well beyond the study of military applications of sonar. India should build a state-of-the-art scientific research institution to study the subject of physical oceanography, chemical oceanography, marine ecology, and marine geology to completely understand the ocean. Indian scientists need to go out to sea and participate in field experiments on a massive scale. The acoustic institutes focused on UDA need to come up with strategic goals.

The Indian Navy is equipped to handle challenges in the underwater domain. The induction of P-8I long-range MPA will provide the Indian Navy with the required capability to keep an eye on the important chokepoints in the Indian Ocean. The 'mission-based deployment' and patrolling of important sea lanes have been prioritized to monitor the undersea environment, after China started actively deploying submarines in the region. However, the surveillance of chokepoints and other important maritime junction in the Indian Ocean remains a major challenge for India. To achieve total situational awareness of the region, India needs to build a network of sensors and sonars in the Indian Ocean to monitor the underwater domain. The wide area of surveillance is a critical area that India needs to specialize in, to mitigate future underwater threats in the Indian Ocean.

Bibliography

21st Century Maritime Security for the Indian Navy. (n.d.). *Boeing India.* Retrieved from: www.boeing.co.in/products-and-services/defense-space-and-security/boeing-defense-space-and-security-in-india/p-8I.page (Accessed on April 5, 2021).

After 29 Years of Service, Navy Bids Adieu to TU-142M. (2017, March 29). *Financial Express*. Retrieved from: www.financialexpress.com/india-news/after-29-years-of-service-navy-bids-adieu-to-tu-142m/607631/ (Accessed on May 12, 2021).

Airborne Tacticians for Navy and Coast Guard. (n.d.). *Indian Navy*. Retrieved from: www.indiannavy.nic.in/content/airborne-tacticians-navy-and-coast-guard (Accessed on May 11, 2021).

Bedi, R. (2020, December 22). India to Commission Second Arihant-Class Submarine in 2021. *Janes*. Retrieved from: www.janes.com/defence-news/news-detail/india-to-commission-second-arihant-class-submarine-in-2021 (Accessed on November 15, 2021).

BEL Inaugurates Upgraded ATDS Maareech Manufacturing Facility. (2020, August 12). *Manufacturing Today*. Retrieved from: www.manufacturingtoday india.com/sectors/8051-bel-inaugurates-atds-maareech-manufacturing-facility (Accessed on October 11, 2021).

Chand, N. (2016, April). India's Anti-Submarine Warfare Capability on the Brink. *SP's Naval Forces*. Retrieved from: www.spsnavalforces.com/story/?id=434 (Accessed on September 12, 2021).

Clark, B. (2015). *The Emerging Era in Undersea Warfare*. Washington: The Center for Strategic and Budgetary Assessments.

Commissioning of Advanced LOFAR Simulator at ASW School. (n.d.). *Indian Navy*. Retrieved from: www.indiannavy.nic.in/content/commissioning-advanced-lofar-simulator-asw-school (Accessed on May 11, 2021).

Defence Acquisition Council, Chaired by Raksha Mantri Shri Rajnath Singh, Approves Capital Procurement for the Services Amounting to Over Rs 22,800 Crore. (2019, November 28). *Press Information Bureau*. Retrieved from: https://pib.gov.in/PressReleaseIframePage.aspx?PRID=1594075 (Accessed on April 4, 2021).

Defence Ministry Issues Formal Tender for Mega Submarine Programme. (2021, July 20). *The Economic Times*. Retrieved from: https://economictimes.india times.com/news/defence/india-issues-tender-for-rs-50000-crore-project-to-build-six-submarines/articleshow/84579400.cms?utm_source=contentofintere (Accessed on November 11, 2021).

De-Induction Of Indian Navy's TU142M Aircraft and Induction of Boeing P 8 I Into INAS 312. (2017, March 29). Retrieved from Press Information Bureau: https://pib.gov.in/PressReleaseIframePage.aspx?PRID=1486005 (Accessed on April 09, 2022).

DRDO. (2017). *Technologies for Underwater Surveillance Systems*. New Delhi: DRDO.

DRDO. (n.d.a). *APSOH*. DRDO. Retrieved from: www.drdo.gov.in/apsoh (Accessed on November 12, 2021).

DRDO. (n.d.b). *USHUS – 2*. DRDO. Retrieved from: www.drdo.gov.in/ushus-2 (Accessed on November 12, 2021).

GRSE Starts Production of Anti-Submarine Ship. (2020, December 31). *Outlook*. Retrieved from: www.outlookindia.com/newsscroll/grse-starts-production-of-anti submarine-ship/2002616 (Accessed on April 15, 2021).

Indian-Designed Data Link II Delivered to Boeing. (2010, May 12). *The Economic Times*. Retrieved from: https://economictimes.indiatimes.com/industry/transportation/airlines-/-aviation/indian-designed-data-link-ii-delivered-to-boeing/articleshow/5921211.cms?from=mdr (Accessed on April 11, 2021).

Integrated Headquarters, Ministry of Defence. (2015). *Ensuring Secure Seas: Indian Maritime Security Strategy*. New Delhi: Integrated Headquarters, Ministry of Defence (Navy).

Kapur, S. P., & McQuilkin, W. C. (2017, February 23). Preparing for the Future Indian Ocean Security Environment. *ORF*. Retrieved from: www.orfonline.org/expert-speak/preparing-for-the-future-indian-ocean-security-environment-chal lenges-and-opportunities-for-the-indian-navy/ (Accessed on July 18, 2021).

Martin, J., Ogle, M., Whalen, D. J., & Wignall, A. (2006). *Multiplying the Effectiveness of Helicopter ASW Sensors*. London: Ultra Electronics.

Mazumdar, M. (2013, August08). *Indian Naval Aviation: ISR Capabilities Set for a Quantum Leap*. Defense Media Network. Retrieved from: www.defensemedia network.com/stories/indian-naval-aviation-isr-capabilities-set-for-a-quantum-leap/ (Accessed on May 11, 2021).

New Mission-Based Deployment Concept to Result in Greater Presence and Visibility in IOR – Admiral Sunil Lanba, C. (2017, October 27). *New Mission-Based Deployment Concept to Result in Greater Presence and Visibility in IOR – Admiral Sunil Lanba, CNS*. PIB. Retrieved from: https://pib.gov.in/PressReleseDetail. aspx?PRID=1507269 (Accessed on October 23, 2021).

PIB. (2020, August 10). *Raksha Mantri Shri Rajnath Singh Launches Modernization/ Up-Gradation of Facilities and New Infrastructure Creation of Defence PSUs and OFB*. PIB. Retrieved from: https://pib.gov.in/PressReleasePage. aspx?PRID=1644892 (Accessed on June 11, 2021).

PTI. (2021, July 17). Indian Navy Gets Two MH-60R Multi-Role Helicopters from U.S. *The Hindu*. Retrieved from: www.thehindu.com/news/national/indian-navy-gets-two-mh-60r-multi-role-helicopters-from-us/article35377299.ece (Accessed on June 1, 2021).

Reim, G. (2021, May 1). India Approved to Buy Six More P-8I Maritime Patrol Aircraft for $2.42bn. *Flight Global*. Retrieved from: www.flightglobal.com/fixed-wing/india-approved-to-buy-six-more-p-8i-maritime-patrol-aircraft-for-242bn/143562.article (Accessed on April 11, 2021).

Scott, R. (2021, March 10). *DRDO Completes Test Milestone for Fuel Cell AIP System*. New Delhi: Navy International.

S-Cube Integrated Submarine Sonar Suite. (n.d.). *Thales*. Retrieved from: www. thalesgroup.com/en/worldwide/defence/s-cube-integrated-submarine-sonar-suite (Accessed on November 15, 2021).

Sharma, R. (2013). Transformation of Indian Naval Aviation Post New Inductions. *Journal of Defence Studies*, 31–48.

Shukla, A. (2014, November 25). Sonar Contract Provides Major Boost to Navy. *Business Standard*. Retrieved from: www.business-standard.com/article/eco nomy-policy/sonar-contract-provides-major-boost-to-navy-114112500024_1. html (Accessed on August 11, 2021).

Shukla, A. (2019, March 7). Indian Navy Signs 10-Year Lease for Third Russian Nuclear-Submarine. *Business Standard*. Retrieved from: www.business-standard. com/article/defence/indian-navy-signs-10-year-lease-for-third-russian-nuclear-submarine-119030701289_1.html (Accessed on August 19, 2021).

SinhA, S. (2021). Decade of the Maritime Air Power. Retrieved from spsnaval-forces.com: https://www.spsnavalforces.com/ebook/75012021.pdf (Accessed on April 09, 2022).

Subramanian, T. (2012, March 23). Navy's Eyes and Ears. *Frontline*.

Thomas, R. (2019, November 26). *Military Logistics Agreements: Wind in the Sails for Indian Navy*. IDSA Comment. Retrieved from: https://idsa.in/idsacomments/military-logistics-agreements-rthomas-261119 (Accessed on September 19, 2021).

Unnithan, S. (2017, December 10). From India Today Magazine: A Peek into India's Top Secret and Costliest Defence Project, Nuclear Submarines. *India Today*. Retrieved from: www.indiatoday.in/magazine/the-big-story/story/20171218-india-ballistic-missile-submarine-k-6-submarine-launched-drdo-1102085-2017-12-10 (Accessed on September 12, 2021).

WESCAM MX™-15, Air Surveillance and Reconnaissance. (n.d.). *L3Harris*. Retrieved from: www.l3harris.com/all-capabilities/wescam-mx-15-air-surveil lance-and-reconnaissance (Accessed on March 11, 2021).

Yoash, R. B. (2016). *Anti-Submarine Warfare Search Models*. Monterey: Naval Postgraduate School, the US Navy.

Zakheim, D. S. (1977). *The U.S. Sea Control Mission: Forces, Capabilities, and Requirements*. Washington: Congressional Budget Office.

6 Conclusion

Unfolding Security Scenario in the Indo-Pacific Ocean

The maritime common in the Indo-Pacific region is becoming more intense in both the surface and subsurface domains. The focus of the discourse here relates to the underwater domain. As in any domain, advances in technology and innovation are the drivers of underwater weaponry with improved capabilities. Better knowledge of the hydrographic and bathymetric conditions of the seas, stealthy and quiet platforms, improved detection techniques, and new weaponry have unleashed challenges for domain awareness, surveillance, protection, and security.

As always, Russia and the United States are trend setters, with China being a close competitor. For example, Janes Navy International reported that in an exercise carried out in the Barents and Norwegian seas in July 2021, unspecified number of nuclear-powered submarines of the Russian Northern Fleet carried out extreme depth exercises. The report goes on to add "crews worked out tasks and tested weapon systems at depths more than 500m" (Jones, 2021). The Russian Yassen/Yassen-M-class of SSN and SSGN submarines typically carry Oniks and Kalibre cruise missiles, and media reports suggest that Russia plans to carry out tests of Tsirkon Hypersonic Cruise Missile from the Yassen-class submarine in November 2021 (Cole, 2021). This missile after qualification can be deployed on Yassen-class submarines. The claimed speed of Tsirkon is Mach 6–8, and at these speeds, it will be a formidable weapon for targeting sea surface and land-based assets. The United States and China also have hypersonic weapons programmes and can be expected to achieve hypersonic missile launch capability from surface vessels and submarines in 3–5 years. Others are playing catchup and the extent of catchup is governed by their naval strength, perceived security threats, and their resource (financial, manpower, training, etc.) situation.

Overall improvements in the endurance as well in stealth features of underwater platforms, in addition to improved onboard weapon systems

DOI: 10.4324/9781003298380-6

and sensors, have helped in enhancing the lethality of conventional and strategic underwater platforms. As a result, even a single submarine can pose a credible threat to a high-value naval surface vessel.

Currently, more than 40 navies operate submarines, and of these, 15 countries are present in the Indo-Pacific region. Among these countries, India and China have both diesel and SSN/SSBN category of Submarine manufacturing capabilities; Japa and South Korea have diesel/diesel with AIP manufacturing capability; and the rest operate acquired diesel submarines. In this milieu, the middle and small countries are recalibrating their naval capabilities, requirements, and acquisition in order to safeguard their maritime sphere. Such a situation has resulted in countries like Pakistan, Iran, and Bangladesh seeking to enhance and update their submarine fleet through, licenced production, or outright purchase. Till recently, Germany, France, and Italy have been the major suppliers of diesel submarines in this region. Russia has exported their Kilo-class submarines to India, and in recent years, China has entered the market.

The other factor affecting the regional maritime security environment relates to the rather volatile situation in the South China Sea. China's Anti-Access/Area-Denial (A2AD) strategy, aggressive posturing against Taiwan, ADIZ declaration, and territorial disputes posing challenging situations. China views the freedom of navigation operations conducted by the US navy and other powers in the SCS and ECS is destabilizing the security environment in this region. The new security pact between Australia, the United States, and UK (AUKUS) brings new dynamic change to the security architecture of the Indo-Pacific. It may contribute to the volatility of the situation as well as augment China's belligerence and A2AD measures. Additionally, the disturbed conditions prevailing in Yemen and Syria, exercise of sanctions against Iran, and piracy have vitiated the security environment in these regions. The prevailing situation has prompted other non-regional powers like the UK, France, Germany, and Russia to also increase their presence and deploy naval assets to secure their energy and trade interests.

The challenges to India in the underwater domain fall under three categories:

1. *Strategic Challenge*: The expansion in Chinese naval presence in the Indian Ocean has been steadily increasing from 2010 onwards. The Chinese BRI and MSR initiatives are directed not only to build connectivity with the rest of Asia and Africa but also with the motive to expand Chinese strategic and defence cooperation with the littoral states of the Indian Ocean. China's interest in the Indian Ocean is twofold: first, to protect the security of SLOCs –during both peacetime and wartime

from enemy blockade or attack – and second, to ensure berthing/basing facility for the Chinese naval warships/submarines through sizeable investments in ports and infrastructure development in the littorals of the Indian Ocean. The Chinese fleet deployment, especially the deployment of its nuclear submarines, aids posturing and power projection. India does not favour this as it challenges its sphere of influence and demands more attention in terms of domain awareness to safeguard against any action adverse to national security.

Additionally, China adopts unequal rules of the seas, which have ramifications for free passage of commercial and military vessels in the South China Sea. For example, China claims the so-called *nine dash line* as its territorial waters with new requirements for vessels passing through these waters. On August 29, 2021, Chinese authorities said that they will require a range of vessels to report their information when passing through its 'territorial waters'. The range of vessels according to the Maritime Safety Administration include submersibles, nuclear vessels, ships carrying radioactive materials, and ships carrying bulk oil, chemicals, liquefied gas, and other toxic and harmful substances (Krishnan, 2021). This dictum is at odds with UNCLOS and the *nine dash lin*e is disputed notably by Philippines, Vietnam, Malaysia, and Indonesia. Also, it is not clear how China plans to enforce this rule. While the rule may be targeting the United States and western fleets mainly, it must be noted that 55% of India's trade passes through these waters and the rule could affect patrolling and surveillance needs of the Indian Navy.

2.   *Military Threat*: Pakistan's submarines and other underwater platforms pose a significant security threat to India. Pakistan is expanding its submarine fleet with the acquisition of the new Yuan-class submarines from China. The newer midget-class submarines Pakistan is presently building indigenously could also be of concern for the safety and security of the west coast of India. Pakistan's Jinnah Naval Base (JNB) in Ormara in Makran Coast has been developed as an addition to the Karachi Port. This will give Pakistan options to shift some of its crucial naval assets to JNB to override possible Karachi port blockade in a hostile situation and prepare for counter-attacks on the Indian naval force. The large construction site in JNB suggests that Pakistan's navy may be building underground facilities to accommodate nuclear weapons or control room for submarines (Panneerselvam & Mukherjee, 2020). Pakistani submarines are equipped with Babur-2 nuclear-capable cruise missile touted as relevant to Pakistan's second-strike capability. These missiles will be kept mated with warhead and in 'ready-to-use' condition unlike their terrestrial-based weapons. The range of Babur-2 is claimed to be

450 km, and with this range, the submarine will be operating in shallow waters not far from the Indian coast and could be a security concern for India.

3. *Technology-Related Challenges:* Newly emerging and disruptive technologies can have significant implications for security. The deployment of UUVs is a prime example in this category. The United States, in particular, has developed a new class of undersea vehicles and systems ranging from 'mini UUVs to fully automated ships'. These platforms can be employed for a wide range of missions in distant enemy waters without human control and can pursue a target or assist the naval ships in detecting and tracking the underwater objects. Obviously, these or similar technologies and tactics will be employed by other navies also, rather sooner, than later. The threat scenario for India will increase manifold when China and Pakistan invest in such technologies and India will have to put into practice appropriate domain awareness tools and incorporate counter-UUV technologies in addition to investing in its own fleet of UUVs for deterrence. China has already progressed in its development of hypersonic weapons. If China fields these weapons on surface ships and submarines, the threat level will increase manifold for India as defences against such weapons are difficult. The security threat could increase further, if hypersonic technology is proliferated and made available to other adversarial countries.

Role of Underwater Domain Awareness for Indian Navy's Anti-Submarine Warfare Operations

The threats India faces is enumerated briefly in the preceding section; it underscores the need for an effective UDA capability. UDA capability should ideally cater to a set of objectives and appropriate strategy to meet the objectives as indicated in Table 6.1.

India has initiated some action for UDA activities and also realizes that more needs to be done. Initial steps in this regard are indicated as follows:

- The Indian Navy possesses multiple ASW platforms, which include surface ships, fixed-wing and rotary-wing aircraft, as well as underwater platforms. UDA functions are served by

  - Hull-mounted or towed array sonars.
  - Ship mounted radars are capable of detecting surfaced submarines and submarines up to periscope depth.
  - Helicopters equipped with ASW sensors operating from warships.

Table 6.1 Indian Navy's Key Objectives, Strategy, and Role of UDA

Role of UDA	• Detect, track, and identify underwater threats/threat situations • Underwater ISR functions • Support security of coastal, territorial, and EEZ underwater assets – both natural and man-made • Safety and rescue functions • Improve situational awareness around critical locations including the crucial straits linking Pacific and Indian Oceans
Objective	• Detect, track, and classify underwater threats • Suppress or destroy adversary threats • Protect India's naval assets (aircraft carriers and SSBNs) • Intelligence functions • Support decision process • Mapping and cataloguing resources in the country's EEZ and securing them
Strategy	• Plan and install adequate and appropriate resources – surface, underwater, and aerial (space, if space-borne detection reaches maturity) • Full-fledged domain awareness using combined functions of platforms/sensors • Intelligence gathering, synthesis/analysis, and cross-check/validation • Modelling, analysis, wargaming, and validation • Research and analysis leading to improvement in sensors and sensing in unmanned systems with higher endurance, higher payload capability, and stealth • Improvements in platform-based weapon systems • Promote research in underwater physics – salinity, acoustic propagation, detection techniques in Indo-Pacific waters • Bilateral and multilateral cooperation, data and intelligence-sharing

Source: Created by author.

• Long-Range Maritime Patrol aircraft like P-8I aircraft with quick
  reaction time – which help reduce the area of probability of submarine presence – and SSNs and conventional submarines – which
  can conduct surveillance in the operational area and create ASW
  cordon in the Area of Interest.
• Other Indian Navy aircraft like IL-38 and Dornier, which can play
  a supporting role.

• The Indian Navy has increased its presence at crucial chokepoints in
  the Indian Ocean under the 'Mission-Based Deployments' programme,
  which provides round-the-clock surveillance of the crucial sea lanes in
  the Indian Ocean.

- Strategic division: sharing of patrol areas with the like-minded countries in the IOR.
- Cooperative information- and intelligence-sharing approach with some regional/major navies is taken up for obtaining better understanding of the maritime environment.
- Jointly and individually carry out frequent and detailed modelling and simulation exercises. This is a necessary adjunct to integrate the UDA information gathered from different sensors and sources.

Recommendations

India has some strengths and capabilities in UDA. This needs to be enhanced manifold. A sustainable roadmap for a mostly self-reliant approach with long-term and short-term goals, is provided in the following recommendations.

1. *Understanding the underwater propagation characteristics:* The tropical climate and the increase in anthropological activities in the Indian Ocean have an effect on acoustic propagation. The rivers draining into the Bay of Bengal – Ganga, Brahmaputra, and Irrawaddy – impact the salinity for some pronounced distance with resultant impact on the acoustic propagation properties. It becomes essential to characterize and demarcate the waters properly and study their influence on acoustic propagation. One can intuitively feel these variations will also have a seasonal dependence, especially relating to the monsoon season extending from June to the end of September. Detailed oceanography and ambient noise-mapping exercises in the Indian Ocean to improve the detection and classification of submarines or any underwater objects are therefore a necessity. Theoretically, directly measuring the ambient noise at depth over a broad spatio-temporal range is a challenging and expensive task (Zhang, Yang, Yang, & Chen, 2019). Moreover, there are inherent limitations in carrying out the task (Das, 2014). Given the enormity of the task, it becomes necessary to plan and prioritize a realizable approach, involving the defence services, research laboratories, and academia as a sustained long-term research initiative.

   Therefore, a proposed collaborative underwater study network involving, agencies like the Navy, the Coast Guard and other security agencies, which comprehend the pulse of the requirements could catalyse a series of brainstorming meetings with stakeholders to hammer out the UDA priorities in the short term. The R&D required to tackle issues in the medium and long terms must be sponsored and a plan of action for technology development must be formulated in pursuit of

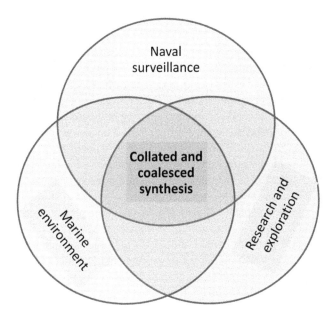

Figure 6.1 Underwater Study Network in India
Source: Created by Author.

that goal. Research priorities need to be linked to infrastructure require-ments, trained researchers and human resource requirements, and most importantly the funding requirements. In such an endeavour, critical planning and fund allotment should be ensured, as subcritical funding will grossly affect the much needed UDA capability and capacity

Finally, a policy paper outlining the approach should be a neces-sary outcome. Such a document will clearly enunciate the vision and research approach and serve as a research driver for the contributing members and it will help indicate to the competitors/adversaries as to where we are headed and how the UDA approach is present to deny, suppress, and thwart any adventurism in our waters.

2. *Increase undersea influence through deploying underwater unmanned systems, seaboard sensors*: Technology plays a lead role in oceanic research as well as in naval operations, particularly in gathering Infor-mation, Surveillance and Reconnaissance (ISR) data. The sonar is a highly effective and widely used technology in ASW operation. They

come in different types: sonobuoys, variable-depth sonar, towed array sonar systems, high-resolution sonar, etc. The Synthetic Aperture Sonar and high-frequency sonar have increased the sonar resolution in detecting underwater objects. Particularly, various research studies show that the Synthetic Aperture Sonars improve the navy's ability to detect sea mines deep under water (Kohntopp, Lehmann, Kraus, & Birk, 2017). The super-computer and improved sonar processing algorithms will aid navy in better target detection and enhance on-board decision making. On the other hand, the growth of unmanned underwater vehicles with the ability to carry different payload can be tailored to engage in the ASW operations. India needs to invest in core technologies like propulsion systems, navigational and communication systems, artificial intelligence/machine learning, collision avoidance systems to make the unmanned vehicle a truly autonomous system. The focus on the core technologies is imperative for India because no country would either share these advanced technologies or offer joint collaboration on building these systems. The public-private model should be explored in the field of development of UUVs and other associated technologies. Investment in the core technologies will allow the Indian private sector to use these technologies for other commercial purposes. Acquiring these technologies not only is important from the ASW perspective but also plays an equally important role in increasing India's influence in the undersea in the Indian Ocean.

3. *Improve situational awareness around the crucial chokepoints in the Indian Ocean:* Actively monitoring chokepoints is crucial to collect intelligence on the movements of submarines or other underwater objects in the Indian Ocean. Given the high density of shipping traffic near chokepoints (Malacca, Sunda, Lombok), deploying underwater sensors at these points is not practical. There are two ways to address this issue: first, to deploy the UUVs or sea gliders routinely along the chokepoints to scan subsurface for any undersea activity of foreign powers and, second, to strengthen India's maritime outreach programme with the littorals. The building acoustic capacity cooperation with regional countries closer to the chokepoints will allow India to observe underwater environment in the region. Programmes like these can help to create greater scientific interaction and establish long-term scientific collaboration with the regional countries in the field of oceanic research, data collection, etc.

The support from countries will be dependent upon geopolitical considerations and in some cases may constrain or negate positive cooperation. Significant diplomatic interaction would be required to get amenable solutions and will have to be a continuous exercise. The

Table 6.2 Strategic Partners: Monitoring Chokepoints in the Indian Ocean

Area	Strategic partners
Straits of Malacca	Indonesia, Malaysia, Japan and the US
Sunda and Lombok Strait	Indonesia, The US, Australia and Japan.
Straits of Hormuz	Iran, UAE, Oman, The US, Japan and NATO navies
Bab-el-Mandeb	Djibouti, Eritrea, Yemen, Japan, France
Mozambique Channel	Madagascar, Mozambique, The US, France

Source: Compiled by author.

benefits of information-sharing towards safety and security of the maritime domain as well as technology-sharing can act as incentives.

The international cooperation in both political and scientific domains to engage in building comprehensive surveillance network in the Indian Ocean can prove deterrent to Chinese submarine operations in the region. The United States and France have military base in the region can network with regional countries to monitor the seafloor for hostile submarine activity. The Quad provides a perfect opportunity for India to build acoustic capacity to counter China in the IOR.

Bibliography

Cole, B. (2021, June 8). Russia to Begin Testing Hypersonic Missile From Nuclear Submarine. *Newsweek*. Retrieved from: www.newsweek.com/russia-testing-tsir kon-hypersonic-missile-nuclear-submarine-weapons-1616816.

Das, A. (2014). New Perspective for Oceanographic Studies in the Indian Ocean Region. *Journal of Defence Studies*, 109–117.

Jones, B. (2021, July 20). *Russian Nuclear Submarines Conduct Joint Extreme Depth Exercises*. IHS Jane's Navy International. Retrieved from: www.janes.com/article/53201/russian-navy-focuses-on-renewal-of-black-sea-submarine-force.

Kohntopp, D., Lehmann, B., Kraus, D., & Birk, A. (2017). Seafloor Classification for Mine Countermeasures Operations Using Synthetic Aperture Sonar Images. In *Oceans 2017 – Aberdeen* (pp. 1–5). Aberdeen: IEEE.

Krishnan, A. (2021, August 30). China to Require Foreign Vessels to Report in Territorial Water. *The Hindu*, p. 11.

Panneerselvam, P., & Mukherjee, A. (2020). *China-Pakistan Economic Cooperation (CPEC) Project: Satellite Imagery Analysis of Port Development at Makran Coast*. Bengaluru: NIAS.

Zhang, Y., Yang, K., Yang, Q., & Chen, C. (2019). Mapping Sea Surface Observations to Spectra of Underwater Ambient Noise Through Self-Organizing Map Method. *The Journal of the Acoustical Society of America*, 146.

Index

Note: Page numbers in *italics* indicate a figure and page numbers in **bold** indicate a table on the corresponding page.

ABHAY 83
acoustic 4–7, 8, 9, 22, 40, 43, 45, 47, 52, 55, 57, 61, 63, 67–68, 75–79, 82, 85–87, 95–96, 98–99
acoustic capability 87
ACTAS 7, 85
ACTUV 28
ADCP moorings 67
ADIZ 92
Agosta 90B 21, 35
AI 7
AIP 10, 15, 16, 22, 23, 24, 26, 27, 33, 36, 37, 54, 80, 84, 87, 89, 92
AMTI 49, 54, 55
Andaman Sea 2, 4, 80, 86
antennas 47
anti-piracy operations 71, 73
anti-ship cruise missiles 16, 26
anti-surface warfare 15, 75
APSOH 83, 84, 88
Arabian Sea 2, 4, 9, 11, 20, 22, 24, 79, 80
Archer (ex-SWE Västergötland) 24
Arctic region 20, 47
area-denial weapons 58
Argentina 38
Argo floats 67
Arihant-Class 79, 88
Armament Package 74
array of sensors 17, 54, 81
Asian rivers 4
ASS 47
Australia's Defence White Paper 2016 25

Babur-3 21–22
Bagamoyo 18
Bahamas 40
Bangladesh 18–19, 24, 26, 35, 92
Barbados 40
Bashi channel 16
bathymetric condition 4, 91
battery 7, 30, 47
Bay of Bengal 2, 4, 9, 20, *24*, 26–27, 35, 80, 96
BEL 7, 85, 88
Bell Labs 43
big data computing 8
biological noise 4
blockade 11, 93
Boeing P8 I 72, 74, 88
Brahmaputra-class 79
brainstorming 96
BRI 3, 92
British Isles 40

Cakra (Type-209/1300) 24
Canadian Navy 62
CASIC Third Academy 16
CCS 11
CDAA 52
C-DAC 80
CETC 51
Challenger (ex-SWE Sjoormen) 24
chemical properties 4
Chernobyl 31
China 2, 3, 7, 9, 12–28, 30–37, 45, 48–56, 84, 87, 91–94, 99

China-Arab Wanfang Co. Ltd.
  (Ningxia) 18
China Communications Construction
  Company Ltd. 18
China Harbour Engineering Company
  Ltd. 18
China International Trust Investment
  Cooperation Group 18
China Merchants Port Holdings Co.
  Ltd. 18
China Overseas Port Holding 18
China Road & Bridge Corporation 18
Chittagong 18
chokepoints 3, 17, 39, 54, 59, 60, 69,
  72, 80, 81, 82, 87, 95, 98, 99
CJ-10 14
classification 60, 66, 84, 96, 99
CMRE 58
CMSI 30
coastal defence 15, 49, 50, 54
coastal waters 5, 7, 12, 38, 48, 67, 73,
  76, 78
collision avoidance 98
COMINT 82
common operational picture 59, 72
CoMPASS electro-optical (EO) pod 76
COMRA 20
contingency 20
COSCO Shipping Ports 18
CSI 64
CSIC 15, 23
CSIR Lab 82

DAC 74
DARPA 45, 55, 62, 69
DASH 45, 55
Datum 77
Decoy 59
defensive ASW 48, 58
Delhi-class 7, 79, 85
detect 7, 8, 27, 30, 37, 40, 43,
  45–47, 54, 57–58, 63, 66, 68–73, 75,
  77–78, 80, 81, 85, 95
detection techniques 7, 91, 95
deterrence capability 12, 21, 79
diesel-submarines 8, 10, 26, 38, 43, 54,
  75, 79, 92
Djibouti 3, 18, 19, 81, 99
DNS 58
dolphin clicks 5
Dornier 228 73, 76

dredging 5
DSRV 68

early warning system 51
earthquakes 5, 51
East Pakistan 11
equatorial current system 4
ESM system 22, 74
Europe 2–3, 37, 39, 40, 58, 65
export control regime 25

Falklands War 38
Fateh 27
Federation States of Micronesia 51
fish vocalizations 5
fleet architecture 28
flotilla 18, 20
French DCNS 23, 25
freshwater 4
fuel cell technology 7, 30, 80

Ganges 4
Gauntlet Pattern 81
Gen(retd) Khalid Kidwai 21
geophysical surveys 5
Ghadir-class 27
GIUK Gap 40, 48, 57
Godavari-class 79
GUGI 46–47, 55
gun running 3, 11
Gwadar 18

HADR 71
HAIG Z-9C 50
Hainan Island 17, 51
Haiyang-class 22 20
Hambantota 18
Hanoi (RUS Varshavyanka)
  24, 26
Harmony-S surveillance system
  47, 56
Hawaii 40, 43
HF radars 52
high attenuations 5
Hokkaido 17
Honshu 17
HSU 001 30
human intelligence 62
HUMSA UG 83, 85
hydrocarbon sensor 75
hydrodynamics 63

hydrographic 4, 18, 20, 28, 52, 91
Hypersonic Missile 91, 94, 99

ICBM 13
IIFP 51
IIT 82
IL-38SD 72–74
Indian Coast Guard 6, 76
Indian navy 6, 7, 11, 53–54, 64, 66,
  68–69, 71–82, 84–90, 93–95
Indian Ocean hydrology 4
Indonesia 2, 19, 20, 22, 24, 26, 33, 54,
  93, 99
Indo-Pacific region 10, 20, 22, 25, 37,
  48, 50, 59, 91, 92
INDUS-NET 86
Industrial Zone of Duqm 18
INSAT communication 67
INS *Brahmaputra* 11
INS Dega 73
INS Garuda 73
INS Hansa 73
INS Rajali 73–74
INS Sagardhwani 86
INS Shikra 73
INS Utkrosh 73
INS Venduruthy 73
INS Vikrant 11
Integrated Data Link 74, 75, 78, 89
intelligence sharing 96
Iran 19, 22, 24, 27, 32, 34, 92, 99
IRBIS 51
Irrawady and Salweeen Rivers 4
Islamabad 21
ISPR 21, 33
ISR 6, 15, 30, 33, 47, 61, 76, 89, 95, 97
ISRO 67, 82
IUSS 43, 45
IUU Fishing 3

Jaish-e-Mohammad 11, 33
Japan 52, 54, 56, 69, 75, 76, 81, 99
Java Sea 2, 68
Jiangdao 49
Jiangkai II 49
Jihadists 63
Jin-class 13, 14, 17
JL-2 missiles 13, 15
JL-3 SLBM 13, 14
JMSDF 3, 17, 20, 52, *53*, 75

JNB 93
JOED 52
Juan de Fuca 31

KA-28PL 50
Kalibr 15, 16, 35
Kalvari-class 79, 80
Kamorta-class 7, 78, 79, 85
Kamov 28 7, 78
Kamov 31 7
Karachi Shipyard 22
Kelvin Hughes 22
Kenya 18–19
Khalid Class 23, 35
Khalifa Port 18
kilo-class submarine 15, 26, 27, 92
Kolkata-class destroyers 7, 78, 79, 85
Kota Kinabalu Naval Base 26
Kuperman 5, 9
Kuril Islands 16
Kyaukpyu 18
Kyushu 17

Lamont-Doherty Earth Observatory 67
land-attack cruise missiles 11
land-based ocean observation stations 65
Larsen & Toubro 80
laser detection 7
LED 7–8
LEMOA 81
LFDS 83
LFVDS 78
LIDR 63
littoral waters 5, 6, 38, 39, 80
LOFAR 43, 76, 88
Losharik AS-12 31
Luyang III 49

MAD 7, 63, 74
Madagascar 19, 99
Makassar Strait 28
Makran: trench 20, 31, 93, 99
Maldives 19, 72, 80
marine scientific research 51
marine vessels 4, 18
Maritime Patrol Radar (MPR) *74*
maritime traffic 2, 3, 54
Mauritius 19, 80
MAWS 74
MDA 4, 6, 8, 14, 62, 63, 64, 66, 68, 74

Meteksan Defence Industries 22
MH-370 68
Mid-Atlantic Ridge 40
midget-class submarines 10, 26, 27, 38, 47, 52, 79, 93
MIHIR–Mid Frequency Airborne Sonar 83
minimal distortion 5
MIT 40
Miyako Strait 16
modelling 95, 96
modernization 4, 11, 13, 15, 21, 22, 26, 35, 89
Mombasa Port 18
monsoon season 4, 96
Monterey Bay Aquarium Research Institute 58
moored buoys 67
MOSFET 86
Mozambique 19, 99
MSR 3, 92
multiple layering 77
Murmansk 47
Myanmar 18, 19, 24, 26

Nabajatra (ex-PRC Ming Type-035G) 24
Nagapasa (Type-209/1400) 24, 26
Nahang 24
narcotics trafficking 3, 9
Nasr-1 (Anti-Ship Cruise missile) 27
NATO 7, 30, 31, 37, 40, 46, 47, 48, 58, 62, 99
naval base: Chinese 3; in Djibouti 3, 19; in Guam 51; in Makran Coast 31; in Sabah 26; Vladivostok 3
naval exercises 71
naval intelligence 14, 25, 56
navigational risk 27
navigation radar 22
nine dash line 93
noise: acoustic 6; ambient 5–6, 8–9, 14, 96, 99; anthropogenic 5; level 14, 77, 79; shipping 5–6; underwater 5; volcanoes 5; wind-generated 5
northern subtropics 4
Nova Scotia 40
NSB 46

nuclear powered submarine 12, 25, 31, 34, 37, 45, 57, 91
nuclear risk 22

Oceanographic Survey Ship 20
ODIS 67
offensive ASW strategy 47–49, 59
oil and gas exploration 5
Olenya Guba 47
Oman 18, 19, 99
Oniks 91
open ocean 4, 6, 12–16, 39, 47, 60, 67, 87
Operation Cactus 72
Operation Parakram 73
Operation Vijay 73
Optronic mast 22, 35
ORBAT 26
organized criminals 11

Pakistan 11–12, 18–19, 20–22, 27, 31, 33, 35, 52, 73, 78, 92–94, 99
PANCHENDRIYA 83, 85
passive sonar 16, 37, 44, 45, 49, 83
Pentagon 56, 63
periscope 22, 35, 94
Persian Gulf 2, 20, 22, 34
Philippine Sea 17
piezo-electric sensors 86
PLA: Navy 3, 4, 8, 12, 16–21, 30, 34, 48–50
PNS *Ghazi* 11
Port Colombo 18–20
port development 18–19, 99
procurement 7–8, 30, 32, 73, 74, 88
profile buoys 65
prototype 40, 55
PTAS 7

quartz sandwich transducer 39

Rajput-class 78, 79, 85
Ras Al-Khair Port Project 18
Red Sea 2
refraction 5
Renhai 49
research vessels 43; scientific 65
reverberation 77
ROK's Cheonan 38

Royal Australian Navy 25
Russian Delta III SSBN 14

SACLANT 40, 58
SAGAR MAITRI 86
Sakhalin 17
salinity 4–5, 8, 66, 86, 95–96
San Francisco State University 58
San Luis 38
Sanya submarine observation network system 51
SATCOM 75
satellite: communication 17; imagery 43; oceanographic 86; remote sensing 65; surveillance 45; tracking 19
SBC 79
Scripps Institute of Oceanography 58
SCSSON 51
S-Cube Integrated 80, 89
sea-based 6, 12, 13, 32, 35, 37, *41*, 46, 79; Nuclear Strike Capability 21–22
seabed 46–47, 51, 55, 58, 62, 65, 70, 71, 81–82, 84
sea bottom 5; observation networks 65
sea-control 35, 82
sea-denial 10, 82
Sea Hunter 28, 30
Sea of Okhotsk 46
sea surface 65–68, 82, 91, 99; buoy arrays 65; temperature 4, 67
Sea Wing 28
second-strike 12, 21, 49, 93
seismic activity 6, 8
Selayar Islands 28
semi-submersibles 65, 69
Seychelles 19, 80
SH-60J Seahawk 52
shallow waters 5, 6, 8, 26, 38, 77, 94
SHARK 45
shipping 2, 5, 6, 17, 18, 20, 27, 54, 63–64, 98
Shishumar-class 79
Shivalik-class 7, 78–79, 85
SIGINT 52
Sikorsky Sea King 7, 73, 78
Sindhughosh-class 85, 26, 79
Singapore 19, 22, 24, 26, 36, 81
situational awareness 1, 62, 63, 64, 65, 71, 75, 82, 87, 95, 98
situ observations 67

SLOC 1–3, 18, 20, 21, 28, 31, 92
SM39 Exocet 23
Snell's law 5
Somali pirates 3
sonar 7, 16, 22, 37–38, 40, 43–45, 47, 49–50, 54, 58, 72, 77, 78, 80; ranger 5; transmissions 5
Sonic System 74, 84
SOSUS 40, *41*, 43, 44–47, 51–52, 69
South Africa 19
South American 20
South Atlantic 38
Southeast Asian 3, 27, 32, 39, 71
South Korea 3, 17, 33, 81
South Sulawesi 28
Soviet Echo – II 31
Soya Strait 17
SPAWAR 58
speed of sound 4, 5, 86
Sri Lanka 18, 19, 52, 80
stakeholders 64, 96
STM 22
Strait: Hormuz 1, *2*, 31, 72, 99; Lombok 2, 3, 17, 19, 20, 54, 75, 98, 99; Malacca 1–3, 17, 19, 20, 54, 72, 75, 80, 86, 98, 99
strategic partners 75, *99*
submarine tender ships 20
SUBTICS *23*
surface vessels 6, 7, 10, 28, 46–47, 49, 72, 84, 91
surface wind 5
SURTASS 43, 52

Tadpole 84
Taiwan 3, 15, 92
Talwar-class 7, 78–79, 85
Tanzania 18, *19*
Taregh (RUS Paltus Project 877EKM) 24, 27
targeting 30, 56, 66, *74*, 91, 93
TCS 61
Tentar Nasional Indonesia-Angkatan Laut (TNI-AL) 25–26
Thailand *19*, 26, 35
Thermax 80
thermocline(s) 4, 7, 9
Tongji University 51
Torpedo *23*, 26, 47, 49, 59, *60*, *84–85*
ToT 7

track 7, 8, 30, 32, 40, 43, 45, 47, 48, 52, 54, 57, 58, 59, 62, 67, 69, 71–72, 74–80, 82
tracking stations 17, 19
transducer technology 39, 82
TRAPS 45
Trigun *74*, *76*
Tsugaru Strait 17
Tsushima 17, 31
Tu-142M 72, 88
Tunku Abdul Rahman (FRA Scorpène) *24*, 26
Type 041 Yuan-class submarines 22
Type 625C Shiyan 20
Type 636A hydrographic survey ship 20
Type-04 SSBN 13
Type-093 13–14, 17
Type-094 13–14, 17, 32
Type-218 SG submarines 26
Type-95 and Type-96 submarine 15–16

U-boat 37, 39
ultrasonics 39, 55
UMS MinyeTheinkhathu *24*, 26
UN Arms Register of Conventional Arms 24
Underwater Great Wall 50, 56
United States 3, 7, 12, 14, 25, 28, 30–31, 37, 39, 40, 43, 45–48, 52, 56, 57–58, 62–65, 69, 73, 75, 81, 91, 92–94, 99
United States Naval Institute (USNI) 58
University of California 58
unmanned: ASW 45; Maritime System *29*; surveillance platform 48; USV 28, 35, 86; vehicle and systems 62, 95, 97
US Homeland Security 63, 69

USHUS *83*, 84, 85, 88
US Los Angeles class SSN 14
USS Indianapolis 59
UUV 7, 10, 12, 28, 45, 69, 98

Varuna MAALAA 86
Vishakhapatnam 11, 68, 69, *73*, 79
VLF station 59

wargaming 95
Wassenaar Arrangements 25, 36
wave-rider buoys 67
weapon delivery 66
Wenz 5, 9
Westland Sea king 7
whale songs 5
Whiskey-class submarines 25
Woods Hole Oceanographic Institution 58
World Trade Centre 63
World War II 5, 31, 37, 39

Xiang Yang Hong 03 20
Xiang Yang Hong 10 20
Xiaoqushan 51

Y-8Q 50
Yellow Sea 17, 38
Yemen 3, *19*, 92, *99*
YJ-12 49
YJ-18 14
YJ-18/B 16
YJ-62 49
Yuan-class submarine *13*, 15–17, 22, *23*, 32, 93
Yugo (DPRK) *24*, 26

Z-18F 50

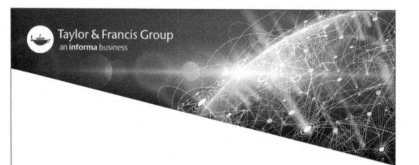

www.ingramcontent.com/pod-product-compliance
Ingram Content Group UK Ltd.
Pitfield, Milton Keynes, MK11 3LW, UK
UKHW020424010325

455677UK00029B/988

9 781032 287577